International Conference on Design, Construction and Operation of LNG/LPG Ships 2017

Glasgow, United Kingdom
29 - 30 November 2017

ISBN: 978-1-5108-8306-2

Printed from e-media with permission by:

Curran Associates, Inc.
57 Morehouse Lane
Red Hook, NY 12571

Printed by Curran Associates, Inc. (2019)

For permission requests, please contact The Royal Institution of Naval Architects at the address below.

The Royal Institution of Naval Architects
8-9 Northumberland Street
London, WC2N 5DA
United Kingdom

Phone: 44 207 235 4622
Fax: 44 207 259 5912

publications@rina.org.uk

Additional copies of this publication are available from:

Curran Associates, Inc.
57 Morehouse Lane
Red Hook, NY 12571 USA
Phone: 845-758-0400
Fax: 845-758-2633
Email: curran@proceedings.com
Web: www.proceedings.com

CONTENTS

* Unavailable at time of going to press

Design, Construction & Operation of LNG/LPG Ships

29-30 November 2017

Glasgow, UK

PAPERS

THE ROYAL INSTITUTION OF NAVAL ARCHITECTS
8-9 Northumberland Street, London, WC2N 5DA, Telephone: +44 (0)20 7235 4622

SHIP DESIGN REVISITED – A MODERN APPROACH FOR A MODERN ERA

D Scott, C Tuzcu, C Pilia, P McMullen and **G Mermiris,** Tritec Marine Ltd, UK

SUMMARY

Over the past forty or so years concerted efforts of the shipping industry, regulatory bodies and academia have resulted in significant improvements of merchant vessels. Modern ships not only satisfy the design specifications (owner's requirements), but also comply with ever increasing numerous and complex regulatory frameworks. However, if the volatility of international markets, and continuous socio-political unrest is taken into consideration, it can be easily seen that although today's "fit-for-purpose" ships cannot be technically faulted, from a financial point of view they perform meaningfully only a fraction of their operational life. This paper will elaborate on the factors and criteria that make a design successful and will demonstrate how the combination of operational experience, state-of-the-art technological developments, and first-principles modelling and simulation can result not only at successful ship designs but also at fit-for-purpose *engineering solutions*.

1. INTRODUCTION

It could be argued that cargo ship design has matured substantially over the past four decades. This could be attributed to continuous development of underlying rules and regulations, concerted industry efforts to reduce accidents and improve safety, and focused R&D activities of the most complex phenomena and processes that govern the construction and operation of ocean going vessels.

It could be further argued that although the shipping industry remains one of the pillars of international economy, it is not immune to economic cycles, and heavy competition. Consequently, a new ship may be designed to comply with all rules and regulations but its performance from a financial point of view is meaningful only during a fraction of its operational life.

So, what makes a design successful these days? Can the compliance to safety and environmental rules, and the deadweight and speed requirements guarantee sustained financial success, and if yes, for how long? If no, then how can this be addressed in a timely and cost-effective manner?

2. OBJECTIVES

Considering that both the ability and appetite of the shipping industry to seek sustainable solutions has been honed to unprecedented levels, it is perhaps time to challenge the overall approach as to what constitutes a good design, rather than continue to focus on the implementation of piecemeal bottom-up technological solutions.

The objective of this paper is to discuss the factors that permeate a new ship design, and the criteria that not only make a design successful, but the timeframe in which the success takes place. The life-cycle benefits of the approach will be demonstrated in the case of 35,000 m^3 gas carrier.

Existing ships and life extension aspects will not be discussed in this paper.

3. SHIPPING, A TRULY INTERNATIONAL INDUSTRY

According to the International Chamber of Shipping, [1], some 90% of the global trade is conducted by 50,000 merchant ships, registered in over 150 nations and manned by over 1 million seafarers. For the first time in history, international seaborne trade surpassed 10 billion tons in 2015, [2].

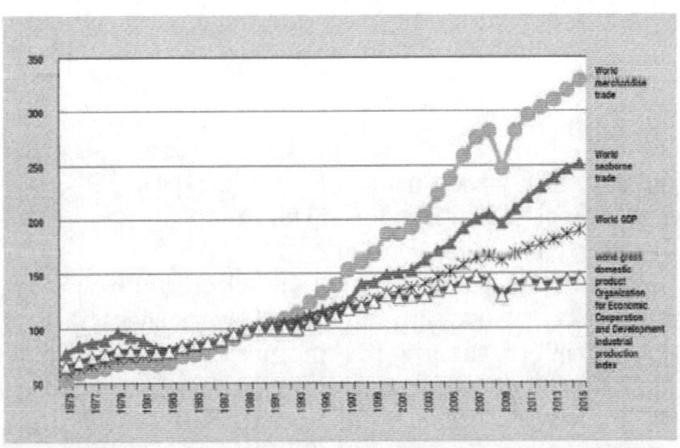

Figure 1: Correlation between global GDP (black line) and seaborne trade (green line) against OECD industrial production index, [3].

With such profound contribution to the international trade activities shipping is not immune to global social, political and economic developments. For example,

- Regulatory developments: Climate change, environmental performance and sustainability have been at the forefront of the global industrial activities for at least two decades. Following the Paris Agreement in December 2015, [3], the shipping industry is expected to face new challenges in the near future, all of which require additional investment and are subjected to strict performance criteria.

- Emergence of large economies: The fact that China and India remain the two fastest developing economies (with growth rates at 6.7 and 7.5 per cent respectively) is well-established. What is less established is that their growth is partially supported by changes in the economic models of developed economies like the United States: from one of the main importers of oil and gas resources, US is slowly changing to an exporter of these commodities. US already exports LNG to some 20 countries, [4].

- Oil price: Crude oil declined some 70 per cent between June 2008 and February 2009, and again some 47 per cent in 2015, and early in 2016 approached the 30 USD/barrel, [2]. The chart in Figure 2 shows the recession periods over the past 30 years, and demonstrates the volatility in the international economy.

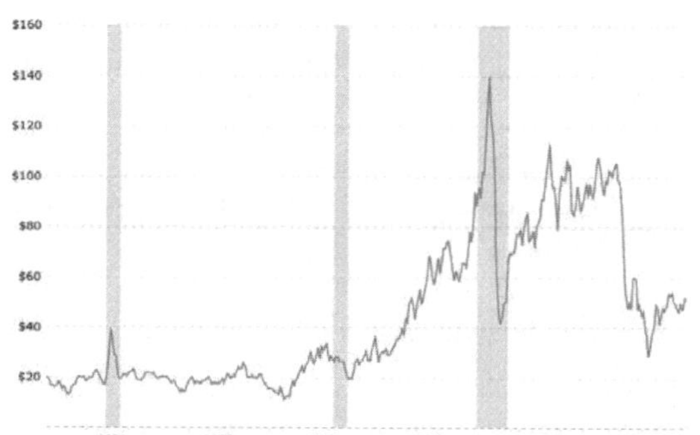

Figure 2: Oil price fluctuation over the past 30 years (shaded areas show recessions), [5].

The rather evident conclusion of the selected few (among many other) points above is that sharp changes in the international economy are taking place over very brief time intervals: the shipping industry is forced to operate and prosper in a rapidly fluctuating market with ever increasing competition and constrains of various natures and on all fronts.

4. INDUSTRY RESPONSE

In response to the current and future challenges the industry invests in and investigates alternative technical solutions to become more cost-effective and competitive. Besides the adoption of state-of-the-aft technologies, more *fit-for-purpose* designs for specific operational requirements naturally dominate the agenda.

4.1 FIT-FOR-PURPOSE DESIGNS

In the context of new assets, the term "fit-for-purpose" balances between commercial and engineering contexts, refers primarily to four dominant attributes of the new ship (in no particular order), and gives ground to a corresponding number of fundamental questions.

- ***Time to market***: Business opportunities of any nature rely heavily on being in the right place at the right time. Currently, new ships require up to two years for delivery, which when considering how rapidly the market fluctuates (Figure 2) seems rather long. *How can the time to market be shortened without compromising the fabrication quality of the new ship?*

- ***Cargo capacity***: The amount of cargo carried onboard reflects commercial requirements, and heavily relies on the 3D shape of the hull envelope and internal layout. In turn, the hull geometry is governed by hydrodynamic, structural, and stability performance criteria, and is constrained by fabrication costs, regulations and logistics. *How can the cargo capacity be maximised without impeding the performance of the ship?*

- ***Maintainability***: Maintenance is perhaps the one operational aspect that is not explicitly addressed during the design and fabrication of a ship, although it will be conducted for some 90 per cent of its life (assuming the standard 25-year life span). Ironically, it is also the most well-predicted one (both in financial and engineering terms) as there is an immense amount of operational experience available in the industry. *How can maintainability be accounted for during design and implemented during fabrication without compromising the functionality of the ship?*

- ***Future proofing***: In a market that is in a constant flux (Figure 2) in financial, regulatory, political and social terms, all future forecasts are dominated by substantial uncertainty both in the required investment and the return on it. Evidently, the financial performance of a ship that is delivered in service today can only be reasonably forecasted in the short term. *What is the most effective way to future proof modern ships?*

4.2 CONTEMPORARY APPROACH TO SHIP DESIGN

Adherence to tradition is a heavily interweaved trait in the culture of the shipping industry, and permeates all aspects of it. Obviously, there is nothing wrong with traditional approaches (and the experience that is built on them) as long as they make a positive contribution towards addressing modern commercial and operational challenges efficiently and cost-effectively.

Ship design is one of those traditional processes, and it starts by the definition of the hull envelope (geometry) in the early stages. The systems for propulsion, cargo handling, docking, refrigeration, crew accommodation, etc., are always addressed a few steps later in the process, as shown in Figure 3. As a result, the onboard systems are simply "made to fit" the available space, with little consideration of the production process, and the efficiency of the ship during its operational life, which corresponds to some 95 per cent of its service life. Practically, this leads to low utilisation of the available space in the engine room, which not only penalises the producibility and maintainability of the ship (high OPEX), but also paves the way for suboptimal determination of its payload capacity.

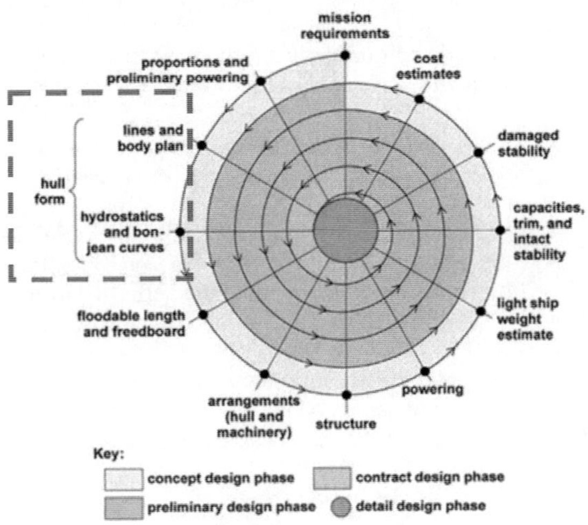

Figure 3: The emphasis of the design spiral is on the definition of the hull form at the beginning of the process, [6].

The result is that disproportionate effort is spent on the performance assessment of the envelope in comparison to the onboard systems. This does not mean that due consideration of the stability, hydrodynamic performance and structural strength should be lessened, but rather that the design of the onboard systems and their spatial configuration should receive at least equal, and in any case more rigorous, attention as early in the design process as possible. After all, the onboard systems will be the main concern of all operations for most of the useful life of the vessel.

The outcomes of this process are ships built to owners' specification and in full compliance with the applicable regulatory framework, yet they are only profitable for a portion of their service lives, or as long as the dominant market factors remain constant.

5. PARADIGM SHIFT IN DESIGN

Notwithstanding the present and future market conditions and a rather obsolete way of defining fit-for-purpose design solutions, a re-think of the overall approach is rather due. The foundations of the proposed approach are equally based on operational experience and state-of-art modelling and simulation.

5.1 OPERATIONAL EXPERIENCE

If there is anything in abundance in the shipping industry, that is experience, and it covers all aspects of operation and management of 87,233 merchant ships as per 2015 statistics, [7].

In the context of this paper, experience can be grouped in terms of

- good operational and design practice (know-how),
- lessons learned, and
- recommendations for further improvement.

Such grouping allows integration of experience into modern designs, and captures potential improvements that can be tested (by numerical modelling and simulation, see Section 5.2) and implemented in future designs.

Unfortunately, the shipping industry suffers from poor knowledge transfer from previous generations of sea farers and ship managers (both at technical and financial levels) to upcoming ones. Thus, integration of historical knowledge in new designs (at least from an engineering point of view) fails in most times and new designs not only look the same but also suffer from the same inherent weaknesses as previous ones.

However, when experience is harmonically coupled with engineering the new designs exceed the focused remits of ships and cross over to the provision of *solutions* with more sustainable and often innovative character.

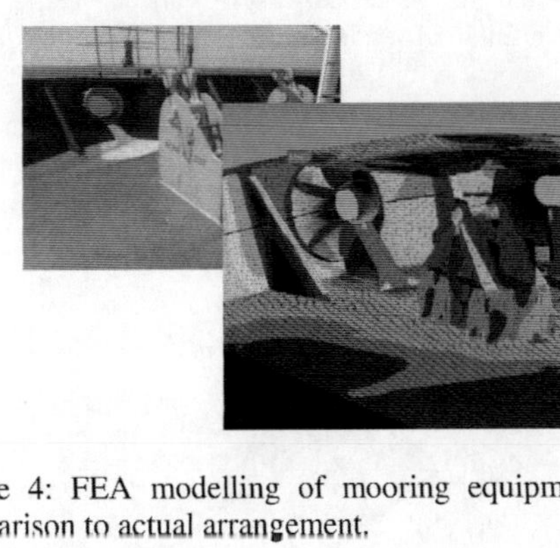

Figure 4: FEA modelling of mooring equipment and comparison to actual arrangement.

5.2 MODELLING AND SIMULATION

Traditional studies in naval architecture often refer to the fact that although the physical phenomena and processes that govern the operation of a floating body are complex (and mathematically challenging to model), yet ships are designed only with basic arithmetic operations like addition and multiplication. This convention cultivates a gap between synthesis (i.e. the basic design development) and analysis. To this day, the latter maintains an exotic character and it is often dismissed as too theoretical or academic in the context of commercial applications, unless regulations explicitly demand use of it.

However, the mathematics governing the stability (both intact and damage), hydrodynamic, and structural performance of a floating body are not only mature but their implementation in commercially available software tools has become the norm. In addition, the continuous improvement of computer technology (processing power and data storage) allows modelling, simulation and assessment to be conducted at any given time, and has the potential to screen out the most promising solutions at minimal cost.

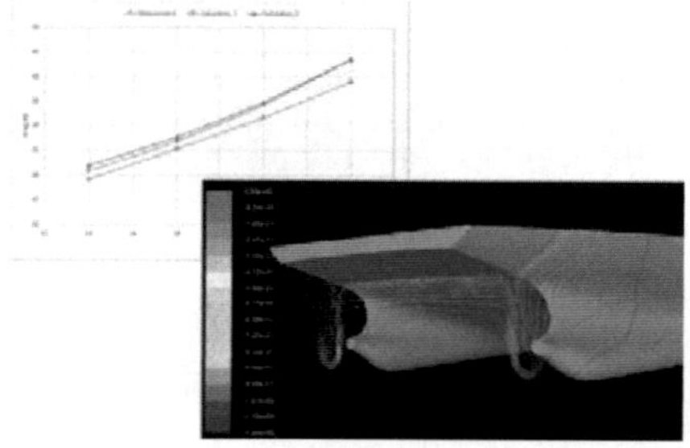

Figure 5: Hydrodynamic assessment with CFD and benchmarking with towing tank results.

As it is often the case, the prerequisite for a successful implementation of the numerical modelling tools and derivation of meaningful results largely depends on the training and experience of the engineering team in charge. When the numerical results can be verified and validated, numerical modelling is a powerful tool during design.

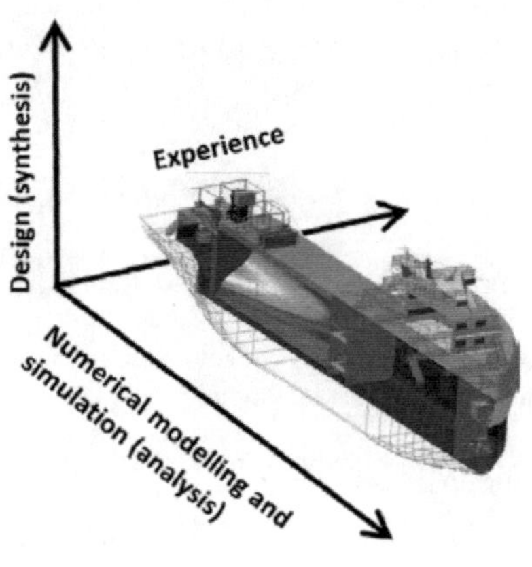

Figure 6: Successful ship designs will result when synthesis, analysis and experience are meaningfully combined.

5.3 BRIDGING THE GAP

Clearly the way forward lies in bridging the gap between synthesis, analysis and historical know-how (Figure 6). This blend can facilitate the identification of the most promising solutions by combining operational experience with the capability to assess performance both on well-known and innovative solutions.

The first step in this direction is to configure a ship as a *system of systems*, [8], from the outset of the design process, thus establishing and assessing the attributes that will permeate all phases of its service life.

In this context, the questions asked in Section 4.1 can now be potentially answered as follows:

- *How can the time to market be shortened without compromising the fabrication quality of the new ship?* This can be achieved by addressing the functions of a ship and defining the independent systems that provide this functionality. Subsequently, the definition of the system interfaces will define how the systems work collectively, and will lead to the selection of optimal arrangements provided that all applicable constrains are met (regulatory aspects, cost-effectiveness, producibility, etc.).

  For example, the power generation system (i.e. the system responsible for generating and distributing power onboard the vessel) and steering system (i.e. the system that provides steering of the vessel) can be combined by the introduction of a diesel-electric system and pods.

  Whether this is an acceptable solution would be argued on the basis of (i) achieving more cargo space (generally diesel-electric systems require less space than diesel-mechanic arrangements), (ii) superior manoeuvrability will require less tug support during docking manoeuvres, and (iii) the whole power generation / propulsion / steering system can be prepared in parallel to the hull fabrication process, and delivered as a single block at the shipyard.

- *How can the cargo capacity be maximised without impeding the performance of the ship?* This question pertains to the continuous effort to maximise the cargo carrying capacity of a vessel for a given envelope (effectively a set of dimensions based on applicable rules and operational constrains, e.g. water depth). Considering that any hull shape can provide only certain amount of buoyancy, the utilisation of every space onboard needs to be re-assessed. To this end, state-of-the-art technological solutions, which require less space and are more efficient (in terms of volume or weight) can be deployed.

Typical examples include the selection of a diesel-electric power generation plant, and the use of compact ballast water management (BWM) systems, which can be installed on the main deck of the vessel (especially when larger cargo vessels are under consideration).

It should be noted that often the applicable regulatory framework can constrain the location of the cargo spaces. For example, cargo tanks on board gas carriers should not be located under the accommodation spaces as per IGC Code, [9]. However, the emergence of SOLAS regulations for *Alternative Design and Arrangements*, and the associated *Risk-based Design* guidelines from Classification Societies, e.g. [10], pave the way for deployment of innovative solutions by de-risking their implementation through combined first-principles simulations and risk-control options identified through expert judgement workshops.

- *How can maintainability be accounted for during design and implemented during fabrication without compromising the functionality of the ship?* Stemming from historical evidence (new building supervision and delivery) and onboard experience, the spatial arrangement of critical spaces like the engine room is configured in detail during the actual assembly process in the shipyard (production dock). The result is a rather convoluted arrangement of pipes that run in all directions and make maintenance during operation difficult.

  Therefore, it should not be surprising that even for newly delivered ships, the engine room arrangement does not match the as-built drawings of the vessel. Moreover, isometric drawings of pipes are at best scarce for ships (contrary to the offshore rigs, for which they are available and required most of the times).

  However, the question remains as to how maintainability can be ensured by design. De-cluttering critical spaces can happen both in terms of well thought-out spatial arrangements (based on operational experience) as well as selection of either smaller number of systems (for example three generator sets as opposed to four) or better distribution of these systems onboard and their interfaces (i.e. with the right number and capacity of switchboards). Systems redundancy can be achieved more cost-effectively if the right systems are positioned in the right place onboard.

- *What is the most effective way to future proof modern ships?* The ever-present question of what to expect (in terms of new regulations, economic recessions, etc.) in the short, medium and long term future is never an easy one to answer. Perhaps embracing this uncertainty as a constraint is the answer to this question (rather than trying to defeat it by piecemeal and short-sighted solutions). As a first step, this can be achieved by aiming for low complexity onboard (in terms of configurations and redundancy levels) from the outset.

  For example, instead of catering for the installation of a particular BWM system (i.e. treatment technology), a space with suitable supports with designated interfaces (suction / discharge) can be reserved. No system needs to be connected (i.e. no purchasing of a system or out-of-service time is required) until the actual implementation date is decided in IMO.

The above aspects can be summarised in a pictorial way with the revised design spiral shown in Figure 7. By addressing the onboard systems selection and configuration early in the process and by building the hull envelope *around* the systems, the end design will be, by definition, more fit-for-purpose. That is, all systems aspects that define most of its service life will be accounted for in equal terms alongside the hull envelope, and the global performance of the ship will be assessed against the owner's specification.

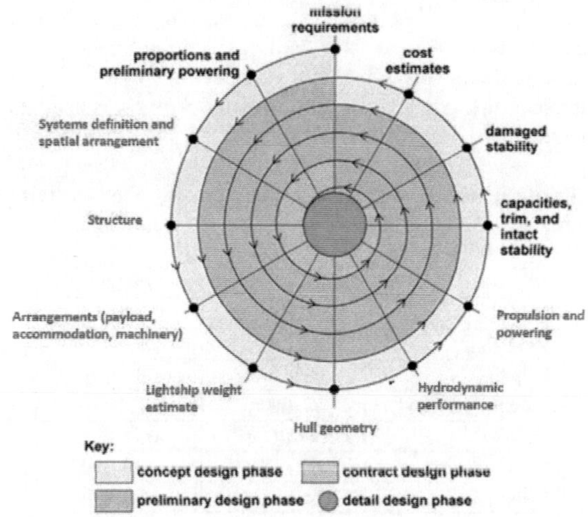

Figure 7: The revised steps of the design spiral.

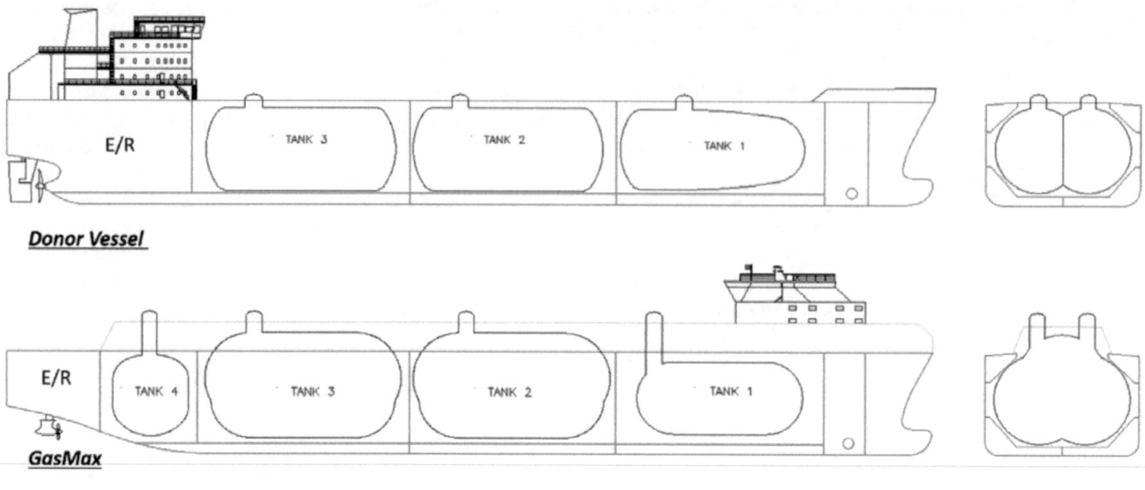

Figure 8: Proposed arrangement of GasMax in comparison to the donor vessel.

6. CASE STUDY

6.1 BACKGROUND AND SCOPE

The technical concepts and ideas, and the status of the wider shipping market and its future challenges presented above are not new. However, from an engineering point of view, the evolution of ship design seems stagnated. The main reason for this is lack of momentum primarily due to insufficient exchange of ideas across the industry.

In order to fuel the discussion of what constitutes a good design these days, the authors of this paper engaged in an internal design project and aimed to demonstrate that without major intervention to the hull envelope of an existing design it is possible to increase the payload of a gas carrier (Table 1) without compromising its seaworthiness or impeding its potentials for future applications. An existing ship (donor vessel) was used as a benchmark for the new design.

The project was named *GasMax*, and it was partially funded by Scottish Enterprise.

	Donor vessel	GasMax
Length	176.00 m	176.00 m
Breadth	29.60 m	29.60 m
Depth	19.40 m	19.40 m
Design Draught	11.00 m	11.40 m
DWT	35,000 m^3	40,000 m^3
Power	7.00 MW	7.00 MW
Speed	16 knots	16 knots

Table 1: Main particulars of donor vessel and GasMax.

6.2 RE-DESIGN OF A 35,000 m^3 GAS CARRIER

The arrangement of GasMax in comparison to the donor vessel is shown in Figure 8. The following points should be noted against the attributes of time-to-market, capacity, maintainability and future proofing (Section 4.1):

- The hull envelope and cargo hold compartments remain the same. The sole exemption is the stern shape (run), which is modified to accommodate the newly introduced pods for propulsion and steering.

- In order to increase the cargo carrying capacity, a trunk is created on the deck. This allowed the introduction of tri-lobe tanks in cargo holds no. 2 and 3. Overall, the cargo capacity has increased by some 15 per cent, and with the introduction of two smaller tanks (no. 1 and 4) the design can be future proofed by introducing partial loading conditions (as opposed to full load or empty ones).

- Diesel-electric power generation and propulsion is introduced. The system is located in the stern of the vessel. For the current version of the GasMax design three generators are accounted for, along with associated peripheral equipment.

- The accommodation block is relocated to the bow of the vessel. Although there are pros and cons for this choice, the intention is to demonstrate that the location of the accommodation block should make the design more fit-for-purpose.

- No. 1 cargo tank is located under the accommodation block. As discussed in Section 5.3, the IGC code prohibits this arrangement in order to safeguard the well-being of the crew against accidental leakage of gas. As such, a risk-based design approach, [10], will be followed, where CFD modelling and simulation of the gas propagation and dispersion will demonstrate how passive and active means, and operational procedures can be used during normal operation or accidental occurrences.

- Finally, it should be stressed that although new systems and arrangements were introduced to GasMax, its compliance with the applicable regulatory framework was maintained.

7. CONCLUSIONS

At a day and age that all market forecasts are governed by volatility and high levels of uncertainty, todays best designs are, in relative terms, not good enough for tomorrow. More than ever before, the shipping industry needs designs that will ensure sustained financial success for the foreseeable future. Unfortunately, this requirement is not commensurate to the state-of-the-art developments in construction, operation and regulation of ships, or the contemporary ship design methodology.

In response to the gap between commercial requirements and available engineering solutions the authors of this paper have demonstrated how efficient design improvements with measurable outcomes can be achieved. The recommendations pertain to the attributes of time-to-market, capacity, maintainability and future proofing, which not only permeate all facets of the operational life of a ship, but also pave the way for the implementation of an ensemble of strategies for more efficient and cost-effective design solutions.

8. DISCLAIMER

The views and opinions expressed in this article are those of the authors and do not reflect the official policy or position of Tritec Marine Ltd, or its parent organisations Northern Marine Group and Stena AB.

9. ACKNOWLEDGEMENTS

The financial support of Scottish Enterprise (contract no.: SPO185923) for the development of the case study presented in this paper is greatly acknowledged and appreciated by the authors.

10. REFERENCES

1. International Chamber of Shipping, http://www.ics-shipping.org/shipping-facts/shipping-and-world-trade, 2017

2. 'Trade and Development Report 2016', *United Nations Conference on Trade and Development (UNCTAD)*, 2016.

3. 'Review of Maritime Transport 2016', *United Nations Conference on Trade and Development (UNCTAD)*, 2016.

4. CLEMENTE, J., 'The Importance of U.S. Oil and Natural Gas Exports', www.forbes.com, 2017.

5. http://www.macrotrends.net/1369/crude-oil-price-history-chart

6. 'Ship Design Process', http://www.marinewiki.org/index.php?title=SHIP_DESIGN

7. Equasis Statistics, 'The world merchant fleet in 2015', www.equasis.org, 2015.

8. Gaspar, H. M., Rhodes, D. H., Ross, A. M. and Erikstad, S. O., 'Addressing Complexity Aspects in Conceptual Ship Design: A Systems Engineering Approach', Journal of Ship Production and Design, 2012

9. IMO, 'International Code for the Construction and Equipment of Ships carrying Liquefied Gases in Bulk', *IGC Code*, 2016.

10. Lloyds Register, 'Risk Based Designs (RBD)', *Ship Right, Design and Construction, Additional Design Procedures*, May 2016.

11. AUTHORS BIOGRAPHY

D Scott holds the position of General Manager / Director at Tritec Marine Ltd. David joined the Northern Marine Group in 2000 and he is a fellow of the Royal Institute of Naval Architects. He has a marine engineering background developing from seagoing positions up to Chief Engineer then leading to fleet management ashore and on to technical consultancy. Significant time shared between operational roles and shipbuilding supervisory roles, at all levels of responsibility, has given David a depth of experience spanning design, construction, operation and disposal of a range of marine assets in the oil, gas and offshore sector. Today, David is responsible for all Tritec Marine's commercial activities and is a keen advocate of exploring innovative technologies and their deployment in commercial applications.

C Tuzcu holds the position of Principal Naval Architect / Stability and Design at Tritec Marine Ltd. Cantekin has a PhD in Naval Architecture from the University of Strathclyde, he is Chartered Engineer with the Engineering Council UK, and member of the Royal Institute of Naval Architects. Cantekin has extensive experience in new ship and marine assets design, as well as conversion and upgrade projects. He is specialised in safety and environmental impact assessment tools and methodology development through focused R&D

projects, and partake in many consultancy projects of various size and complexity. As a formal MCA surveyor, he has plan approval experience, and has been involved in regulatory development work for various IMO instruments. He is a member of the international Ship Stability R&D committee.

C Pilia holds the position of Senior Naval Architect / Structures at Tritec Marine Ltd. Cristian has an MSc in Naval Architecture and Marine Engineering from the University of Trieste, he is a Chartered Engineer with the Engineering Council UK, and a member of the Royal Institute of Naval Architects. His main area of technical expertise is structural analysis (modelling and simulation) of fixed and floating assets. He has been involved in numerous projects entailing both global and local structural modelling and simulation.

P McMullen holds the position of Principal CAD Engineer at Tritec Marine Ltd. Peter has an extensive Structural Design background (2D / 3D and measurement surveys) with over 30 years' experience within the Marine and Offshore Industries. Peter began his career with the Ministry of Defence in the Design Division, creating Structural Design Drawings for the Royal Navy. Subsequently, Peter specialised on ship conversions for private companies, and spent the last 8 years in the offshore Industry working on various FPSO and offshore installation contracts.

G Mermiris holds the position of Technical Manager at Tritec Marine Ltd. George joined Tritec in March 2015. He has an MEng and PhD in Naval Architecture from the University of Strathclyde, and he is a Chartered Engineer and member of the Royal Institute of Naval Architects (RINA). George has extensive experience in leading teams (often multi-disciplinary) and projects of various size and complexity. George's primary expertise lies within the field of numerical modelling of marine structures, and he has been involved in projects covering a variety of topics related to design and operation of ships and offshore structures (seakeeping, stability, mooring analysis, manoeuvring, risk modelling, fire engineering and passenger evacuation).

NEW GENERATION OF MULTIPURPOSE GAS CARRIERS

C Guerrero, Bureau Veritas Marine and Offshore SAS, France

SUMMARY

Ethylene and Ethane has been usually transported in relatively small ships of less than 20,000 m^3 equipped with semi-refrigerated type C tanks. However, new technical developments and the evolution of the regulations are now making possible the utilization of larger ships with innovative containment systems such as bi-lobular type C, independent prismatic and membrane. On the other hand, restrictive air pollution regulations are driving engines manufacturers to develop innovative technologies considering new marine fuels. In fact, ethane gas can offer similar advantages than methane as a fuel in marine combustion engines. As a consequence some propulsion solutions are evolving to respond to a potential new demand.

NOMENCLATURE

LEG	Liquefied Ethane or Ethylene Gas
LNG	Liquefied Natural Gas
VLEC	Very large ethane carrier
EGR	Exhaust Recycling System
SCR	Selective Catalytic Reactor
NOx	Nitrogen Oxides
SOx	Sulphur Oxides
CO2	Carbon Dioxide
IMO	International Maritime Organization
IGC Code	International Code for the Construction and Equipment of Ships Carrying Liquefied Gases in Bulk
MN	Methane number
DF	Dual fuel
HFO	Heavy fuel oil
MDO	Marine diesel oil

1. INTRODUCTION

Commercial developments are behind any technical innovation. On this respect, USA is positioned as ethane exporter as a result of the incipient shale gas industry.

Routes from North America to North Europe, Far East and potentially to South America are expected to increase significantly in the coming years.

In addition ethane is considered as an environmental friendly solution also for power generation as NOx, SOx and particulate matters are reduced significantly to similar levels than using methane as fuel.

Thus ethane transportation is expected to grow and the use of ethane as a fuel is becoming a new trend, not only in power generation on shore but also as a clean transportation fuel for ships, i.e. in compliance with IMO regulations mentioned in MARPOL Annex VI.

Classification societies are involved in the approval of cargo containment systems used for large scale transportation of liquefied ethane in accordance with international and class regulations. They also assess main engines technologies such as dual fuel engines able to burn new fuels like ethane on board of ships. However, as mentioned in the revised IMO IGC code (International Code for the Construction and Equipment of Ships Carrying Liquefied Gases in Bulk) the use of ethane as fuel might be only considered provided that there is the same level of safety as with natural gas as fuel.

Some innovative designs of multipurpose gas carriers have been also developed following the evolution of the liquefied gas industry. These LNG/LEG/LPG carriers are multi product ships and specific issues will have to be considered in the approval.

2. THE CONTAINMENT OF THE CARGO

2.1 LIQUEFIED GASES BEHAVIOUR

In order to assess a new technology for liquefied ethane or the necessary modifications to be made in an existing technology already used for other liquefied gases we must identify first of all the characteristics of every product able to be transported by sea. These characteristics have to be analysed for gas and liquid forms to assess the risks and prevent any accident.

In fact ethane is liquid at atmospheric pressure when it is cooled at -89 Celsius. This means that a containment system has to provide necessary insulation to prevent excess of boil off but much less that the required insulation for LNG which is kept cool at -161 Celsius at ambient pressure.

On the other hand, density of liquefied ethane is higher than LNG. As a consequence for the same filling level in a cargo tank the weight of the cargo is also higher when liquefied ethane is stored as compared to LNG and sloshing pressures on the containment system more important.

In relation with potential loss of containment it is perfectly described in the IMO IGC Code the necessary materials to be used for the construction of the tanks with reference to typical transportation temperatures at atmospheric pressure of -55, -90 and -165 Celsius for LPG, liquefied ethane and

LNG respectively among others. In particular for the cargo piping system, stress analysis is required in case the cargo temperature is -110 Celsius or below. With reference to the steel structure of the ship temperature distribution around the tanks will give necessary requirements for the steel plates to be used.

In addition to the above, the characteristics of the gas resulting from the evaporation of the liquefied gas are also relevant for the purpose of preventing any accident to happen. In particular the density of the gas and the flammability must be considered in case there is a leak of liquefied gas or vent on board of the ship (see figure 1).

Properties	Methane	Ethane
Boiling point, 1 bar [deg.C]	-161,5	-88,63
Density, gas 1 bar, 0 deg.C [kg/m3]	0,717	1,35
Density, saturated liquid 1 bar [kg/m3]	422	540
Critical pressure [barg]	46,41	47,83
Critical temperature [deg.C]	-82,45	32,3
Latent heat of vaporization [kJ/kg]	511	490
Flammability range % in oxygen	5.4–17%	3.0%–12.5%
Lower calorific value MJ/kg	50.1	47.6
Lower calorific value MJ/m3	35.9	64.4

Figure 1.

The vapours clouds of the liquefied ethane being slightly heavier than air will possibly remain in the space affected the range of flammability being similar to methane.

2.2 TRANSPORT OF LIQUEFIED GAS – TYPE OF SHIPS

LEG may be considered in between LPG and LNG for transportation purposes taking into consideration the density of the liquefied cargo and the temperature of transportation (please refer to figure 1).

Because of the characteristics of the gas, ships for transport of LEG are similar to LNG carriers as LEG are basically transported at atmospheric pressure and low temperature in fully refrigerated ships. However, the size of the LEG carriers has been traditionally small as compared with fully refrigerated LPG or LNG carriers thus traditional containment systems for LEG have been based on the independent cylindrical type C type.

Figure 2 – Bilobe tanks under construction for a series of 27,000 m^3 LNG/LEG/LPG carriers for Evergas.

As mentioned by the IMO IGC Code, temperature of the cargo and type of tank govern the requirement to have full, partial or secondary barrier. For type C tanks secondary barrier is not required whether for other type of tanks and transportation temperatures below -55 Celsius, corresponding to LEG, secondary barrier is mandatory, full or partial in case of independent type B tanks (see figure 2).

As volumes are expected to grow in LEG transportation by sea, type B or membrane technologies may be also used so necessary assessment is required.

2.3 CONTAINMENT SYSTEMS FOR VLEC

The first system proposed for a liquefied gas carrier was in fact based on the membrane technology Mark III developed by GTT that was first designed for LNG. Already in 2014 several class societies issued an approval for the transportation of other liquefied gases in this system based on the same principle, i.e. fully refrigerated transportation at atmospheric pressure (please see figure 3 & 4).

Figure 3 – Inside view of Mark III tank for a standard LNG carrier of 160,000 m^3

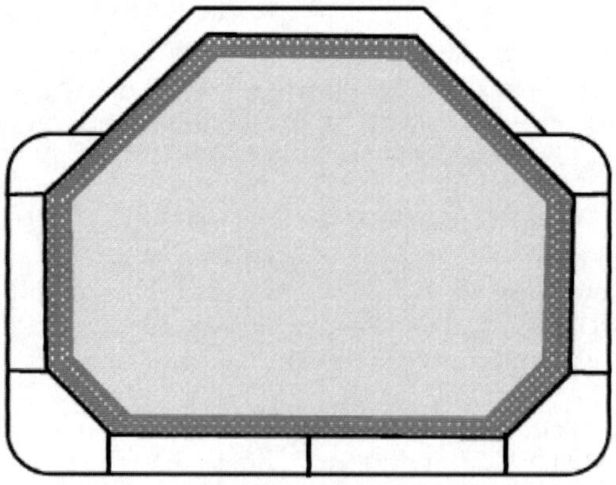

Figure 4 –Typical section of membrane system showing the membrane insulation and inter-barrier spaces.

As mentioned in the previous paragraph 2.1. some technical verifications are necessary due to different factors in comparison to LNG such as:

- Different chemical composition of the liquefied gas
- Higher density of the liquid cargo
- Higher density of the vapours
- Different boiling temperature

Typical filling levels of 10% or below or 70% or above are permitted for this membrane technologies when filled with LNG for a liquid density of 0,47 t/m^3. Being the density of the LEG 0,54 t/m3 it is required to assess the membrane strength for the increased sloshing pressures caused.

In fact, a reassessment of the membrane system is required due to the heavier cargo to demonstrate that the system can withstand the pressures and forces generated in a static and dynamic condition. Fatigue assessment has to be conducted as well to demonstrate that the Mark III is suitable for the LEG cargo (see figure 5).

Figure 5 – Sloshing studies with a Mark III system scale 1:1 using water in Marin tank test facility.

On the other hand, and not related to the membrane system itself, the cargo system should be also revised and dimensioned in accordance with the new cargo, i.e. instrumentation (e.g. level switches, level gauges), pumps, compressors and heat exchangers or vaporisers which are to be designed accordingly.

As mentioned, this innovative application is the adaptation of a proven design for LNG containment at atmospheric pressure. The system keeps the same insulation thickness (primary and secondary spaces) as a result the boil off of the liquefied ethane can be considered as basically half of what the Mark III grants for LNG. Anyhow, there should be some advantages of using reliquefaction systems on board (proven systems are already installed in small LEG carriers) or burn the ethane as fuel in main engines or other onboard equipment (this aspect will be developed in more detail in the following chapter).

Further to the necessary studies developed to certify Mark III for the transportation of LEG (and eventually LPG), GTT decided to develop a new system specific for LPG based on a single barrier membrane system considering that the transportation temperature is not below -55 Celsius.

Obviously additional systems already developed for LNG and approved by IMO such a type A or B may be modified in the future to transport liquefied ethane as well.

Figure 6 – Independent tanks. Type B SPB IHI system used for LNG transportation (above) and Type A used for LPG transportation (below).

3. THE USE OF ETHANE AS MARINE FUEL

3.1 REGULATIONS

Stringent regulations are proposed to reduce NOx, SOx (and particles) and CO_2 in different locations and new regions will joint emission control areas for instance the Baltic and North Sea will become a tier III area with regards to NOx in 2021. As far as SOx is concerned IMO also confirmed recently the world cap of 0,5% sulphur content in marine fuels from 2020. Other SOx reduction regulations are implemented other regions such as California, China or Hong Kong. Last but not least CO_2 emission will have to be monitored in accordance to the EU MRV and IMO regulations.

Alternatives for the compliance may be basically systems reducing NOx in exhaust gas such as SCR or EGR. On the other hand SOx and particles may be removed by cleaning systems such as different type of scrubbers (open, close loop, hybrid or eventually dry systems).

However, for liquefied gas carriers, the extensive experience related to the use of boil off gas in power generation on board (using the LNG vapours) will drive the stakeholders to consider clean fuels such as ethane or propane as marine fuel.
Chapter 16 of the revised IGC Code permits the use of other cargo vapours as fuel for liquefied gas carriers.

Nevertheless, necessary evaluation of risks has to be carried out to demonstrate equivalent level of safety than other installations using methane as fuel, if acceptable to the Administration.

3.2 ETHANE BURNING ENGINES

As of today two different technologies are proposed for marine applications as follows.

The 2 stroke MAN ME-GIE is based on the principle of gas injection and diesel cycle which requires high pressure. Unfortunately at the time of development of the largest ethane carriers ever built by SHI, the development of the technology was not in a final stage so the engine has not yet installed in any ship. It is however foreseen the installation of an EGR system to comply with tier III NOx regulations.

The 4 stroke DF Wärtsilä is based on the LNG fired 4 stroke engine with the necessary modifications to cope with the different characteristic of the ethane gas. The engine meets the IMO Tier III regulations without need of after treatment device while using either LNG or LEG as fuel. The engines have the capability to seamlessly switch between LNG, LEG, HFO or MDO without the need for any modifications to hardware and with uninterrupted operation, thereby setting a new standard in fuel flexibility. The first engines in service are onboard of the 8 ship series owner by Evergas which are example of modern mid-large scale LEG carriers.

3.3 ETHANE AS FUEL. MAIN CHALLENGES

Generally speaking manufacturers of DF and gas engines use the so-called MN for specification of gas quality requirements. MN characterizes the tendency of gas spark ignition engine and the knock resistance. The composition of the gas and in particular ethane composition in a 4 stroke engine may lead to a knocking issue as a pre-ignition phenomenon can happen in a 4 stroke dual-fuel engine (based on the Otto cycle) because of the low methane number as compared to boil off gas of LNG (see figure 7).

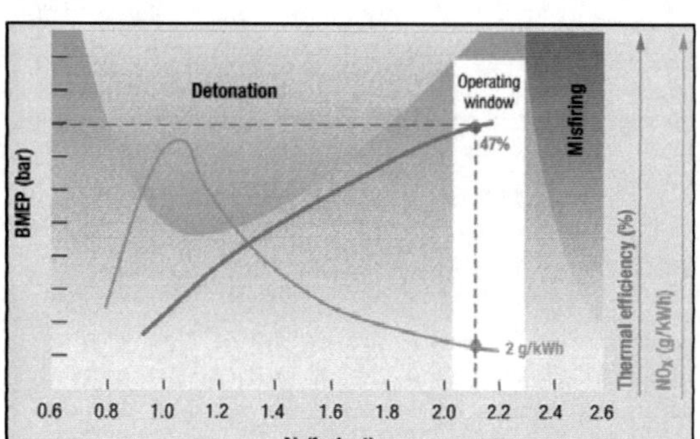

Figure 7 – (Courtesy of Wartsila) Operating window to prevent Knocking or Misfiring.

Similar behaviour of a 4-stroke Otto cycle engine can happen when using heavier hydrocarbon gases like propane as fuel because of the high methane number of the gas. On the contrary, if nitrogen content on the gas is significant, misfiring effect may occur.

As previously described a gas leak inside the engine room can lead to a flammable atmosphere and may cause fire or explosion when a source of ignition is nearby. In a typical configuration of ventilated double pipe (gas safe machinery space) a gas leak outside the external pipe can then lead to a significant accident.

3.4 ETHANE AS FUEL. SPECIFIC SOLUTIONS

The first ethane as marine fuel application was developed for the series of 8 x 27,500 m3 LNG/LEG/LPG carriers was initiated by the owner (Evergas) early in 2015 when the first ship was already launched and the 2 main 4 stroke 6L50DF Wartsila engines for propulsion installed on board. At that time the engines were already designed and certified for the use of LNG vapours as fuel.

A first meeting was organized in BV in order to review the main challenges brought by the change of gas fuel. It was first determined the list of applicable rules for this project as follows:

- Bureau Veritas NR 467 - Rules for the Classification of Steel Ships
- IGC Code Chapter 16 Article 16.9

From the point of view of the regulations the issue of a gas leak was immediately identified during a risk analysis. The outcome of the analysis led to modification in the engine room ventilation (air flow and position of intakes and out-takes) and changes in the engine room gas detection system. The most appropriate gas detection devices, capable to detect both Methane and Ethane mixtures had been selected (only for the installation running with both fuels). In addition, due to the case of ethane gas fuel systems, additional gas detectors had been fitted where gas may accumulate including low down (lower zone) in the vicinity of the bilge wells.

In addition, to assure the engine capabilities modifications including a lower compression ratio, different turbocharger nozzles and de-rating of the engine were required.

With regards to the engines de-rate the engines were originally designed to burn gases having a methane number higher than 70. Pure ethane is characterized by a MN of 43 which may lead to a maximum theoretical power output of 63% of maximum continuous rating (MCR) to avoid having the risk of knocking. Nevertheless, since this output was deemed to be too low to guarantee normal vessel operations, Wärtsilä decided to re-configure the engine and tune the combustion parameters to mitigate the self-detonation problem and increase the available power of the engines.

After testing and validating the new configuration, the target of attaining 73% MCR while operating on ethane was achieved which has been considered acceptable for the intended commercial purpose of the ships. In addition, the DF engines have the capability to switch between LEG to MDO or HFO without interruption of operation.

An additional potential issue was identified by the engine manufacturer with regards to the absorption capability of the lubricating oil used. The absorption capability is higher with ethane than with methane as ethane is heavier than methane. However, taking into account previous engine manufacturer experience with LPG fuelled engines for shore applications, and being propane and butane gases even heavier than ethane, the same oil specification has been considered valid without any problem on absorption noted.

Last but not least, following the IMO procedures, the owner had to seek the agreement of flag administrations for ethane as a fuel as ethane was not mentioned in the previous IGC code. A relevant justification file was prepared by BV and presented to the relevant Flags for acceptance. As a result, the 8 LEG carriers are now trading between USA and North Europe using cargo vapours as fuel and showing to the shipping industry that clean and efficient alternatives to low sulphur fuels are reliable and commercially feasible.

4. CONCLUSIONS

The number of requests for approval of new ship technologies linked to commercial developments is growing. Specifically liquefied gas carrier designs have significantly evolved in the last years with the latest commercial initiatives linked to large distribution of LPG and LEG and increased tons x mile.

Significant technical developments have taken place recently as far as cargo containment systems and marine engines are concerned as a response to an incipient new trend for large scale transportation of liquefied ethane by ship.

The international and class regulations do not necessarily cover some technical evolutions of relevant ships equipment. In particular brand new systems have been proposed in relation with cargo and power generation on board of new generation ethane carriers.

The classification societies are helping the industry to assess these new technologies by reviewing the designs, surveying the construction and participating into the necessary tests.

5. AUTHORS BIOGRAPHY

Mr. Carlos Guerrero is involved in the business development of Gas Carriers and Oil Tankers in Bureau Veritas head Office in Paris since 2012. His main role is the field of oil and gas, following closely different subjects such as innovation in gas carrier designs, new liquefied gas containment and propulsion systems, the use of LNG as a fuel, etc.

Before joining BV, Mr. Guerrero has been largely involved in shipbuilding activities in several Spanish shipyards, working in the production department, technical office, and finally in the commercial department. Carlos Guerrero has also ship operational experience because of his previous position as commercial manager of Union Naval Valencia, S.A. shipyard belonging to Spanish shipping company Boluda Group.

Mr. Guerrero is M.Sc. Naval Architect and Oceanic Engineering by the Universidad Politécnica de Madrid (Spain), where he finalised the thesis in March 1998 and is member of Spanish Official College of Naval Architects.

LNG BUNKERING SHIPS

C Guerrero, Bureau Veritas Marine and Offshore SAS, France

SUMMARY

LNG as a marine fuel is taking a position as one of the more interesting solutions to comply with stringent environmental regulations for ships. Global sulphur cap of 0,5% establish recently by the IMO will probably drive shipowners to consider clean fuels instead of cleaning systems eventually for bigger ships involved in world trade. Small scale LNG carriers with ability to perform LNG bunker operations have become nowadays a clear new trend in LNG shipping to respond to an expected growing demand for LNG as marine fuel. Bureau Veritas was first involved in the certification of small scale LNG carriers more than ten years ago and has become a leader these days in the certification of LNG bunkering ships. Among others Bureau Veritas developed in 2015 specific regulations for the assessment of LNG bunker ships. The regulation NR.620 covers safety issues and is currently the instrument to grant the service notation LNG bunkering ship.

NOMENCLATURE

AIP	Approval in Principle
Barg	Pressure in bar of the tank gauging
BOG	Boil off gas
CCS	Cargo Containment System
DA	Design assessment
DE	Diesel electric
DF	Dual fuel
ERS	Emergency realise system
GASA	General Approval for ship application
GCU	Gas combustion units
GTT	Gaztransport & Technigaz
HFO	Heavy fuel oil
IGC Code	International Code for the Construction and Equipment of Ships Carrying Liquefied Gases in Bulk
IMO	International Maritime Organization
ISO	International Standard Organization
LNG	Liquefied natural gas
MARVS	Maximum allowable relief valve setting

1. INTRODUCTION

Sophisticated LNG carriers for break bulk distribution commonly known as small scale LNG carriers have been developed for many years already. The first modern designs are in service since 2000. This market segment represent less than one per cent of the total LNG shipping capacity but it is still a high value fleet taking into account the unitary cost of this type of ships.

However the concept of small scale LNG carrier has slightly changed in the past few years due to new developments linked to the distribution of LNG as a bunker fuel.

LNG as a marine fuel is step by step taking a relevant position as one quite interesting solution to comply with stringent environmental regulations for ships. The global sulphur cap in marine fuels established by the International Maritime Organization will possibly drive shipowners to consider clean fuels instead of exhaust gas abatement systems in the horizon of 2020.

As a consequence additional LNG bunker supply capacity will be required to satisfy the demand and therefore it is expected that new small scale carriers dedicated to deliver this bunker fuel will be deployed in the most important bunkering hubs all over the world.

Several small scale LNG ships below 10,000 cubic meters have recently entered into service. These ships are more innovative than standard small LNG carriers and have been designed and built to perform break bulk LNG distribution but mainly distribution of LNG as a bunker fuel.

Specific issues have to be considered during the assessment and validation of the ships design and during the construction or conversion process at the shipyard.

Among others, the transfer system and the issues in relation with the boil off gas management are the most important aspects to be considered.

2. THE MODERN SHIP DESIGNS

2.1 SHIPS SIZE

The LNG bunkering industry for the time being seems to follow the path of the HFO bunker maritime industry.

A typical size for an oil bunkering tanker is in the range of 5.000 dwt, representing a cargo volume of approximately 5.000 m^3. In the case we switch to LNG as fuel we will need to approximately double this cargo volume to provide the same autonomy.

However, it remains to be seen if LNG bunkering ships over 10.000 m^3 will be design as ports and terminals of operation may have restrictions related to manoeuvrability for instance and the risk of operating larger ships in commercial ports would be increased.

Ship's dimensions are also linked to the type of CCS selected as we will describe in detail in the following paragraph.

2.2 CONTAINMENT SYSTEM

There is a variety of LNG containment systems that can be used in accordance to the main international regulation, the IMO IGC Code. The Classification societies have their own regulations for the assessment of CCS where safety aspects are also based on the IMO IGC Code.

Type C tanks have been widely used in small scale LNG carriers followed by membrane systems (GTT). However only two units in the range of 20.000 m3 have been built with Type B Moss type as described in the figure 1 below.

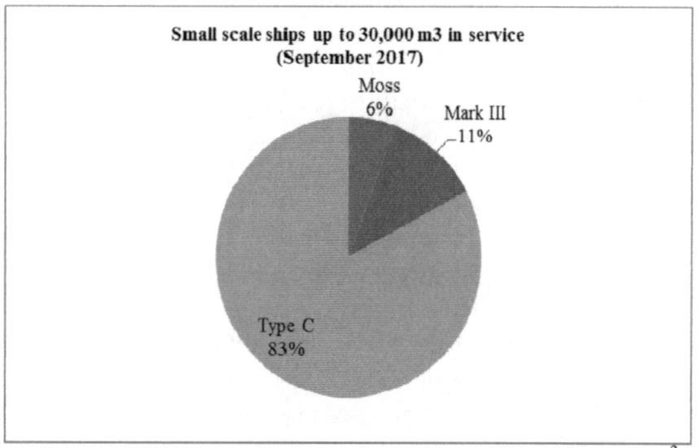

Figure 1. Containment systems for ships up to 30,000m^3

Prismatic tanks (independent or membrane) are preferred to optimize the cargo capacity or in other words, the ship's dimensions, because the tanks are adjusted to the ship's hull in such a way that void spaces surrounding the tanks are almost negligible (see figure 2).

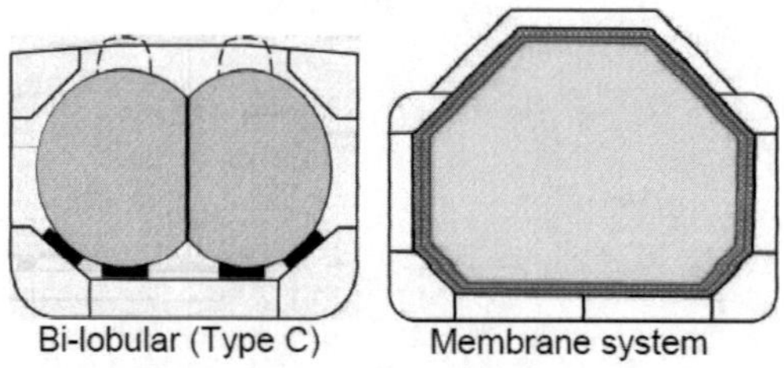

Figure 2. Prismatic tanks cross section

It has to be noted anyway that following the IGC Code CCS which require a secondary membrane need to be insulated from machinery spaces by means of a cofferdam. In addition, a cofferdam with heating is usually considered in membrane ships as a transversal division between tanks to optimize the steel grade of the structure.

Having said the above, the volume optimization provided by the membrane and prismatic CCS has further advantage for applications on board of large ships.

LNG bunkering ships designers have preferred type C tanks because of the ability to withstand pressure inside the tank, generally up to 5 barg. As the LNG is boiling and the insulation is not perfect, the heat ingress the tank and vapours are generated leading to a pressure rise. One of the ways to handle the BOG is by pressure accumulation in the tank. This will be described in more detail in the paragraph 3 of this document. A pressurized CCS implies reinforcement in the tank structure that gives additional weight.

Consequently another positive aspect for membrane tanks is the weight of the tank as compared to a pressurized type C. Membrane CCS are designed in accordance to the IGC Code to withstand a maximum pressure of 0,7 barg. The reduction of tank weight may lead to a reduced lightship, very convenient for shallow draft designs or to an increase of cargo capacity for the same draft and dimensions.

In addition to the conventional systems already described, other innovative systems have been recently developed. As requested by the designers, a two tier approval process is offered for the new systems to facilitate marketing the products. In any event classification society and IMO IGC code regulations will apply as a minimum set of requirements.

AIP is proposed in the first instance in order to ascertain that there are no showstoppers to implement the CCS on board of a ship. The scope of an AIP is not pre-determined and can easily be developed with a limited amount of documents to be mutually agreed between the classification society and the client.

A more detailed assessment called DA (or GASA) may be performed later when all necessary detailed information is available. A DA approved containment system can be finally installed on board of the ship in compliance with the applicable regulations above mentioned.

Among the systems recently developed a new membrane system and a type A tank have been recently introduced to the industry.

Another interesting application had been developed by GTT with the aim to cope with pressure rise in the tank for LNG bunkering applications. It is the membrane Mark III system with 2 barg MARVS which is giving additional flexibility for LNG bunkering ship applications. An AIP has been granted based on paragraph "Equivalents" of IGC Code chapter 1, and is not yet implemented in any ship.

2.3 PROPULSION SYSTEM

Basically DF or gas engines have been proposed for the LNG bunkering ships to produce energy onboard for propulsion and electricity generation.

The technology is widely available nowadays and different types of engines have been approved such as 4-stroke DF, 4-stroke spark ignited (gas engines) and lately 2-stroke DF.

Different combinations of propulsion can be also used by means of mechanically or electrically driven propellers. DE installations can be also considered giving additional flexibility to the system in combination with azimuthal propellers for instance.

The most popular system however is based on 4-stroke propulsion DF engine(s) coupled to propeller(s) via reduction gear and several DF generator sets connected to a main electrical distribution switchboard. A shaft generator could be also an option.

An alternative that gives flexibility is a DE plant based on several DF generator sets all connected to the main electrical distribution switchboard and two electrical motors driving azimuthal propellers. The advantages of the system might be the increased manoeuvrability due to the propellers, flexibility and reduction of overall running hours for the engines meaning also less maintenance cost. Anyhow, from the CAPEX point of view this solution may be expensive.

A new solution has been proposed already by means of 2-stroke DF engines that will normally drive at low speed the propeller without the need of a reduction gear in between. The BOG consumers in this case will be completed by the necessary 4-stroke DF generator sets to produce the necessary electricity onboard.

The operational profile of the ship and the area where the ship is to perform bunkering activities are key points for the selection of the propulsion system. A highly variable work load of the engines may be a reason to consider eventually an electric hybrid system to store the electricity produced when burning BOG in batteries and use the energy during peak consumption periods.

2.4 COOLING/RELIQUEFACTION SYSTEM

Subcooling of cargo or reliquefaction of BOG may be a good method to keep the cargo cool and reduce the pressure in the tank.

Full reliquefaction systems on board of standard LNG carriers were designed and installed in the Qatar fleet of Q-Max and Q-Flex ships for instance. These ships had a very specific requirement to reliquefy all the amount of BOG to save the cargo and keep it cool at any time.

However, reliquefaction systems are a challenge. The design of this equipment has to take into consideration motions, such as rolling, pitching, etc., which have never been design considerations for LNG plant on shore. In addition, reliquefaction systems are complex and expensive systems based on cryogenic moving components such as piston or centrifugal compressors. Reliquefaction systems are based on the principle of a circuit where nitrogen (or mixed refrigerant) as a cooling media is compressed, then expanded and finally circulated through a heat exchanger where compressed BOG is circulated through another circuit. The BOG is condensed when passing through the refrigerant heat exchanger and finally sent back to the cargo tank.

Subcooling of cargo is another method used to reduce temperature and therefore the pressure inside the tank. In this case the liquid cargo is pumped to a heat exchanger where the nitrogen loop cools down the cargo.

A recently implemented subcooling system proposes to extract the BOG and circulate into the heat exchanger as compared to the standard reliquefaction systems where the BOG is first of all compressed before being sent for condensation to the heat exchanger.

2.5 TRANSFER SYSTEM

The world's first commercial STS transfer of LNG was performed in the Gulf of Mexico by *Exmar* and *Excelerate* in 2005. Nowadays the system is certified in accordance to international standards for the transfer of LNG and approved by Class societies.

The STS operation may be carried out in protected or open waters and has been proven in a wide range of meteocean conditions. Nowadays STS with flexible hoses is widely accepted in the LNG industry and over a thousand of STS have been already performed.

Specific LNG STS transfer procedures were developed by the shipping companies involved and reviewed by the class societies.

Basically two different bunkering systems have been developed.

The first solution is based on flexible hoses combined with other systems, such as cranes, davits or tailor made structures to handle the hoses and connections. This is typically the system used in the marine bunker industry.

Rigid transfer systems similar to the already proven terminal loading arms have been also designed and installed in at least one bunkering ship. The system may be less flexible than the flexible hose solution but may be more reliable.

In any event, the safety of the system must be kept and necessary assessment, including tests will have to be properly done.

Based on all the above accumulated experience Bureau Veritas developed in October of 2015 the Rule Note NR.620. The requirements set out in the regulation cover special arrangements for ships carrying LNG which will transfer that LNG to ships using LNG as fuel.

In particular, this regulation covers the design and installation of the LNG transfer systems within Section 4 from the bunkering ship to the receiving ship and the vapour transfer system from the receiving ship to bunkering ship, including LNG hoses, transfer arms and auxiliary systems for handling the LNG system.
Another important aspect of the NR.620 is the requirement to perform a risk assessment of the design and the installation of the LNG transfer system.

The purpose of the risk analysis is to assess the consequences of a failure affecting the concerned systems, an LNG leakage, a tank over-pressurization or any other risk leading to safety issues. The required analysis can be a HAZOP analysis or another type of analysis providing equivalent information for the LNG transfer system.

The required analysis is to be based on the single failure concept, which means that only one failure needs to be considered at the same time. Both detectable and non-detectable failures are to be considered. Consequences failures, i.e. failures of any component directly caused by a single failure of another component, are also to be considered.

3. THE CHALLENGES

3.1 HANDLING OF BOG

Methane resulting from the evaporation of the LNG will harm the atmosphere because of the green-house gas properties of this gas, 25 times more than CO2. IMO IGC Code does not allow venting of cargo vapours because of this reason and the risk to have a flammable atmosphere.

As mentioned in the IGC Code there are different methods to be used in a gas carrier in order to keep the pressure and temperature in the tank below the setting of the MARVS. The different ways described are as follows:

- Reliquefaction of cargo vapours;
- thermal oxidation of vapours;
- pressure accumulation; and
- liquid cargo cooling.

The key issue when dealing with the BOG on LNG bunkering ships is the different profile of operation as compared to a small scale LNG carrier.

First of all we have to consider that a small scale LNG carrier is designed to load the cargo in a terminal, transport the LNG at a normal sailing speed and finally discharge the cargo in another terminal. During the transportation, BOG generated is burnt in the engines on board and used mainly to propel the ship and for the electrical consumers. Depending on the required power during this phase a part of all BOG is consumed.

LNG bunkering ships will normally have less BOG consumption in the DF or gas engines onboard. The reason being basically the way the ship may be operated as longer stand-by periods are foreseen for this type of ships. In addition, LNG transfer to gas fuelled ships may lead to an increased amount of cargo vapours since there might be a vapour return connection to the LNG bunkering ship to reduce the bunkering time. Then, BOG to be taken into consideration is much more that the amount considered in a small scale ship purely designed for LNG transportation.

As mentioned in paragraph 2.4. new small scale units for reliquefaction or cargo subcooling have been developed. These systems have been installed already in big LNG carriers but the technology is evolving and some LNG bunkering ships have also installed these systems already. Nevertheless the efficiency and reliability of small scale installations remain to be seen as it is a very new technology.

As mentioned in the IGC Code, thermal oxidation method means a system where the boil-off vapours are utilized as fuel for shipboard use or as a waste heat system (the so called GCU). As described in paragraph 2.3 different technologies are developed and can be used to reduce the pressure in the tanks. Anyhow, it is expected that the consumption of BOG is reduced on an LNG bunkering ship as compared to a standard unit.

The control of temperature and pressure by accumulation of pressure is basically related to the selection of CCS. As described above a pressure of 5 barg can be typically considered for type C tanks so additional amount of BOG can be stored for a period of time that is to be calculated and justified by the CCS designer. As an example of holding time please see figure 3 where depending on the cargo tank construction material a duration of 49 days is calculated until the tank reaches 3,5 barg pressure.

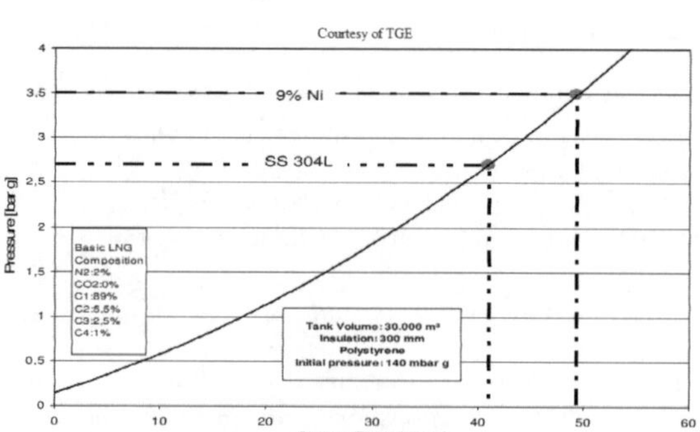

Figure 3 – Holding time for a 30,000 m3 tank.

As a consequence, pressurized tanks will give additional flexibility to the ship BOG handling onboard. However, having a warm LNG inside the cargo tank does not necessarily help for the typical operations to be carried out by a bunkering ship. In figure 4 it is shown the relation between pressure and temperature for LNG. As pressure increases, the LNG becomes warmer, as a consequence the density of the LNG decreases and the calorific content of the fuel per volume unit is reduced. Warm LNG as a fuel will not contain the same energy than a cool fuel and increased BOG amount will be generated during the bunkering operations.

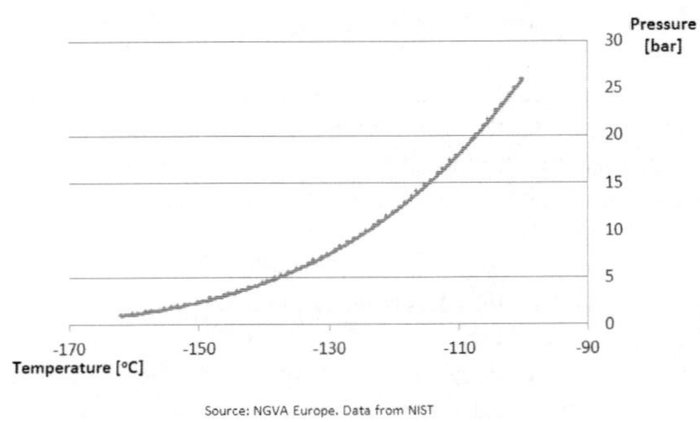

Figure 4 – Temperature vs. Pressure for LNG.

3.2 ASSESSMENT OF TRANSFER SYSTEM

LNG is transported at temperature of around -160 Celsius at atmospheric pressure and at around -150 Celsius in pressurized type C tanks. As a cryogenic liquid, an LNG spill will cause brittle fractures in the carbon steel leading to cracks which may affect the overall strength of the ship (see figure 5).

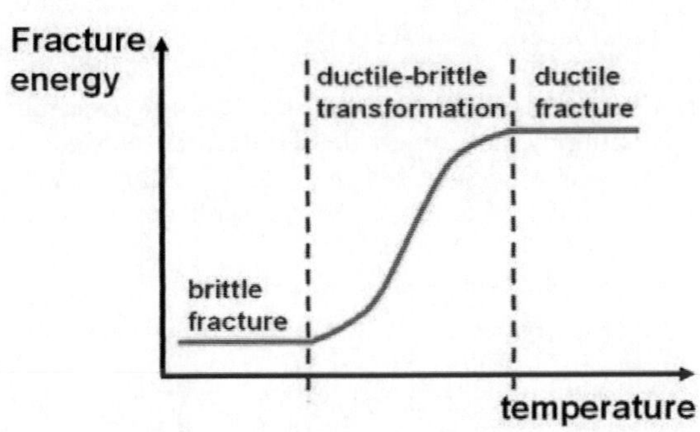

Figure 5 – Energy needed to fracture the material related to temperature.

In addition, once evaporated, escaping LNG becomes a low flash point gas flammable in a composition of between 5 and 15 percent in air at atmospheric conditions and with consequential risk of fire if an ignition source is nearby. See figure 6.

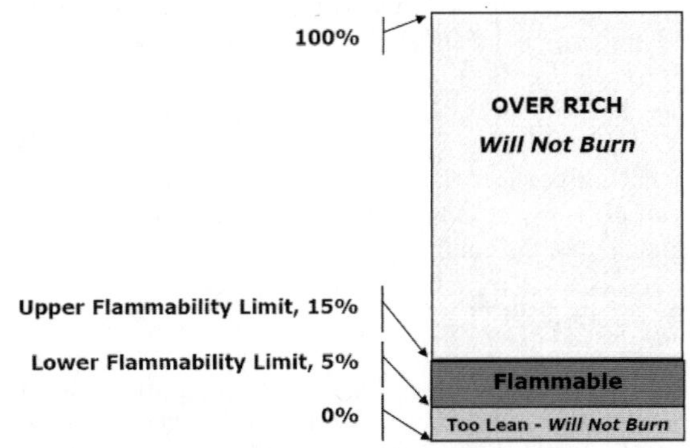

Figure 6 – Flammability range of methane in air.

The components considered in the LNG transfer systems must be designed and manufacturer to avoid LNG spills which may have dangerous consequences.

Different regulations as the NR.620 described above have been developed to assess the transfer systems. On the other hand, standardization is a key aspect in the industry, since LNG fuel supply will not be regional and standard components will be preferable worldwide. For this reason, industry and standardization bodies have been involved in working groups to elaborate the required standards for the different components of a transfer system.

The ISO published the ISO/TS 18683:2015 covering "Guidelines for systems and installations for supply of LNG as fuel to ships" which also refers to the EN 1474-3 standard.

In addition, ISO has recently developed the international standard 20519 "Ships and marine technology — Specification for bunkering of liquefied natural gas fuelled vessels" which has been published last February. The working group involved in this ISO standard has included a wide spectrum of specialists from the maritime industry, equipment manufacturers, the Society for Gas as a Marine Fuel (SGMF), trading companies, class societies, international registries and the U.S. Coast Guard. In this document it is stated that all the components of the transfer system shall be fabricated to meet or exceed the applicable sections of the engineering standards, the IGC/IGF Codes, in addition to other requirements listed in this document. In particular, elements such as hoses, swivel joints, flanges, bearings, ERS, breakaway couplings, transfer arms and other transfer system will have to be in compliance with EN, BS, ISO and ASME relevant standards.

One of the standards considered in the above mentioned ISO 20519 is the ISO 16904:2016 which actually addresses several issues related to the transfer arms such as the definition of the length and the configuration of the arms, arms description, design basis, swivel joints, structural bearings, accessories, pipework and fitting, welding, corrosion and embrittlement protection and maintenance.

Last but not least there is a new draft document ISO/AWI 21593 - Marine LNG fuel bunkering quick connect/disconnect coupling standard (QCDC) still under discussion by means of which a minimum set of standards for the couplings will be proposed.

3.3 OPERATIONS

An LNG bunkering ship under operation in a commercial port involves risks that have to be identified. A leak of LNG or vent of natural gas may lead to an escalation of events which would impair other port operations.

Traditionally LNG carriers operate between terminals specially designed for this type of ships. Otherwise, LNG bunkering ships will be operated in other type of terminals or ports, including oil, chemical or passenger terminals then involving a more challenging operation profile.

Another important aspect to be taken into consideration is the possibility to perform LNG transfer operations simultaneously with cargo operations (including passengers transit in case of a cruise or ferry ship).

As a consequence, Port and local authorities, supported by experts are developing the necessary HAZID studies to put in place recommendations aiming to eliminate risks during the operations of the ship. Filters will be considered including ship transits and LNG bunkering procedures.

Some of the outcomes of the risk analysis in way of recommendations will have to be incorporated in the ship design as well.

As described above, in particular for the transfer system, a Hazop is requested by Bureau Veritas and the outcomes of the study will have to be considered in the system design.

In addition to the risk identification, port authorities and other industry stakeholders have put in place guidelines for bunkering in order to consider all necessary aspects involving LNG bunkering.

4. CONCLUSIONS

A careful assessment of all parts of the LNG fuel supply chain is required to avoid accidents.

The LNG bunkering ships are becoming a key component of the delivery process considering the growing demand in the coming years. Consequently, necessary regulations have to be in place to insure that no leaks, vents or other issues occur during bunkering operations or ship transits leading to significant accidents.

Classification societies have been deeply involved in the certification of LNG carriers but specific considerations applied to LNG bunkering ships and in particular to the transfer system.

Different technologies are available to insure the compliance with the regulations from classification societies and IMO and commercial considerations will also take place to help the designers to propose a better solution for a specific application.

Standardization of transfer systems and in particular couplings for worldwide operation is key together with a proper identification of the risk involving the ship and its operations to prevent accidents from happening.

5. AUTHORS BIOGRAPHY

Mr. Carlos Guerrero is involved in the business development of Gas Carriers and Oil Tankers in Bureau Veritas head Office in Paris since 2012. His main role is the field of oil and gas, following closely different subjects such as innovation in gas carrier designs, new liquefied gas containment and propulsion systems, the use of LNG as a fuel, etc.

Before joining BV, Mr. Guerrero has been largely involved in shipbuilding activities in several Spanish shipyards, working in the production department, technical office, and finally in the commercial department. Carlos Guerrero has also ship operational experience because of his previous position as commercial manager of Union Naval Valencia, S.A. shipyard belonging to Spanish shipping company Boluda Group.

Mr. Guerrero is M.Sc. Naval Architect and Oceanic Engineering by the Universidad Politécnica de Madrid (Spain), where he finalised the thesis in March 1998 and is member of Spanish Official College of Naval Architects.

PURPOSE DESIGNED LNG BUNKERING VESSELS (LBV): SOME CHARACTERISTICS AND FEATURES

W J Choi, PhD (Eng), Lloyd's Register Asia, **J M Navarro**, Lloyd's Register EMEA and **R D Tustin**, BSc (Hons), MRINA, CEng, Lloyd's Register EMEA

SUMMARY

For ocean going ships where operational endurance, and range, requires larger capacity bunker fuel tanks the adoption of gas as a fuel has established a need for a new type of dedicated LNG Bunkering Vessel (LBV): a combination of a small scale LNG carrier with specialist LNG systems for handling the Ship to Ship (STS) LNG bunker transfer. Three early examples of purpose designed LNG Bunkering Vessels (LBV) have now entered service in 2017 to support LNG bunkering operations in European Ports.

Comparing LNG Bunkering Vessels (LBV) in service, under construction and in design development, and drawing on Lloyd's Register knowledge from involvement in a number of these early project examples, the paper will offer an overview of characteristics and features and outline some considerations for specification, design and operation of a dedicated LNG Bunkering Vessel (LBV).

NOMENCLATURE

BOG	Boil-Off Gas
BOR	Boil-Off Rate
CCS	Cargo Containment System
CPP	Controllable Pitch Propeller
DF	Dual Fuel
DFDE	Dual Fuel Diesel Electric
ECA	Emission control area
ESD	Emergency Shut-Down
FPP	Fixed Pitch Propeller
GCU	Gas Combustion Unit
HFO	Heavy Fuel Oil
IMO	International Maritime Organization
LNG	Liquefied Natural Gas
LBB	LNG Bunkering Barge
LBV	LNG Bunkering Vessel
MARPOL	IMO Marine Pollution Convention
NOx	Nitrogen Oxides
NECA	NOx emission control area
SECA	SOx emission control area
SGMF	Society for Gas as a Marine Fuel
SOx	Sulphur Dioxide
STS	Ship to Ship
TPS	Terminal to Ship via Pipeline
TTS	Truck to Ship
ULCS	Ultra Large Container Ship

1. INTRODUCTION

Implementation of stricter limits of emissions of sulphur dioxide (SOx) in Annex VI of the MARPOL Convention, as well as regional SOx emission control areas (SECAs), is encouraging adoption of methane gas, stored as a liquefied natural gas (LNG) bunker, as a fuel for commercial shipping. To meet the need for LNG bunkering of gas fuelled shipping a new type of purpose designed LNG Bunkering Vessel (LBV) has entered service in 2017.

1.1 PAPER OBJECTIVE

The objective, or purpose, of this paper is to offer:

- An overview of some design characteristics and features of LNG bunkering vessels (LBV)
- Outline some considerations for specification and design of an LNG Bunkering Vessel (LBV)

2. ADOPTION OF GAS AS A MARINE FUEL

From a June 2017 consolidation of data (IHS Fairplay & Clarksons, 2017) for the world fleet of commercial ships (excluding LNG carriers and LNGC derivatives using cargo Boil-Off Gas (BOG) as a fuel) there are ninety-eight (98) gas fuelled ships in service and thirty-four (34) on order. Of particular note is that the order book (1.5 m GT) of gas fuelled ships is equal, in gross tonnage, to the fleet of ships in service. It may be concluded that larger gas fuelled ships are being ordered for employment for deep sea voyages and international trade and this conclusion is supported by the ship types in the order book of gas fuelled ship.

Latest data published by the Society for Gas as a Marine Fuel (SGMF) (2017) indicates, when compared with June 2017 consolidation (IHS Fairplay & Clarksons, 2017), a larger fleet of gas as a fuel ships with one hundred and nineteen (119) in service and one hundred and seventeen (117) ordered.

Differences arising in a comparison of 2017 consolidation (IHS Fairplay & Clarksons, 2017) and SGMF data (2017) needs detailed study, however, it is expected that the differences will be accounted for by:

- Ship size and type limitations in respective data sets
- Inclusion of cases where gas as a fuel machinery is installed in so-called *"gas ready"* ships where space is allocated at build for future fitting of an LNG bunker tank and gas fuel supply system

2.1 LNG SUPPLY: PORT OUTLOOK

In a survey (Benito, 2014) of the outlook of twenty-six (26) ports on possible provision of LNG bunkering facilities more than three-quarters of ports surveyed had indicated that they expected to supply, by 2020, LNG as a fuel bunker for gas fuelled ships. Ship to ship transfer from LNG bunkering barges (LBB) or vessels (LBV) was seen as preferred method of LNG bunker delivery by these ports.

2.2 METHODS OF LNG BUNKER TRANSFER

LNG transfers in bunkering operations are expected by one of the following methods;

- Ship to Ship (STS)
- Truck to Ship (TTS)
- Terminal to Ship via Pipeline (TPS)

In practice the method adopted for LNG bunker transfer is dependent on LNG bunkering facilities within the port, the amount of LNG to be transferred, and availability of locations for LNG transfer that will ensure uninterrupted, and unimpeded, port operations when LBV is double banked moored for STS LNG transfer to a receiving ship.

Figure 1: Double banked mooring for LNG bunkering *(Courtesy of Sirius Shipping)*

In a 2013 (Danish Maritime Authority, 2012) study the following recommendations for bunkering methods were made based on the bunkering volumes required by receiving vessels:

- Ship to Ship (STS) transfer is recommended for receiving vessels requiring more than 100 m^3 LNG
- Truck to Ship (TTS) transfer is recommended for receiving vessels requiring up to 200 m^3 LNG
- LNG Terminal to Ship via Pipeline (TPS) is recommended for all bunkering volumes where there are recurrent customers (i.e. receiving vessels) and space available for bunkering facilities

For Ship to Ship (STS) bunkering the study (Danish Maritime Authority, 2012) highlighted that one LNG Bunker Vessel (LBV) per receiving vessel is appropriate to minimize port turnaround time which implies matching of LNG LBV cargo to the bunker tank capacity of the receiving ship.

3. LBV: A NEW SHIP TYPE IN 2017

A new type of purpose designed LNG Bunkering Vessel (LBV) has emerged to meet the need for LNG bunkering operations: a combination of a small scale LNG carrier with specialist equipment for handling and Ship to Ship (STS) LNG bunker cargo transfer.

Three purpose built LNG Bunkering Vessels (LBV) have entered service in North European waters in 2017 with initial deployment in the port of Antwerp, port of Rotterdam, and for harbour and coastal service off Scandinavia and in the Baltic Sea. Each ship is positioned close to an LNG bunker supply source but could be expected to sail to other ports and locations to supply LNG bunker fuel.

These first examples of LNG Bunkering Vessels (LBV) are shown in Figure 2 photographs. In Table 1 particulars of these first LBV examples are compared.

Top: ENGIE Zeebrugge (courtesy of ENGIE)
Middle: Coralius (courtesy of Royal Bodewes)
Bottom: Cardissa (courtesy of Shell International Ltd)

Figure 2: Photographs of first examples of LBV.

Table 1: Comparison of particulars of purpose built LNG Bunkering Vessel (LBV) entering service in 2017

LNG Bunkering Vessel	***ENGIE Zeebrugge***	***Coralius***	***Cardissa***
Cargo capacity (m^3)	5100	5800	6500
Designer	Hanjin Heavy Industry & Construction Co. Ltd, Busan, Republic of Korea	FKAB, Uddevalla, Sweden	STX Offshore & Shipbuilding, Jinhae, Republic of Korea
Shipbuilder		Royal Bodewes, Hoogezand, The Netherlands	
Yard/NB/Hull No.	268	803	4080
Classification	Bureau Veritas	Bureau Veritas	Lloyd's Register
LNG Bunkering Operation			
LNG Terminal	Fluxys, Zeebrugge	SkanGas, Brofjorden	Gate, Rotterdam
Owner	Gas4Sea	Sirius Veder Gas AB	Shell
Charter Party	ENGIE, Mitsubishi Corporation, NYK	SkanGas	Shell
1st LNG Bunkering (date/ship)	14-06-2017/*Auto Eco*	19-09-2017/*Fure West*	-
Main Dimensions			
Length over all, LOA (m)	107.6	99.7	119.9
Length btw. perp., LBP (m)	101.8	92.2	113.1
Breadth moulded, B (m)	18.40	18.0	19.4
Depth moulded, D (m)	9.00	12.1	10.9
Design draught, Td (m)	4.7	5.7	5.8
Deadweight and Tonnage			
Deadweight (tonnes) (Td)	2990	3100	4300
Gross tonnage	7403	6105	9816
LNG Cargo Tanks			
Cargo tank & system design	TGE-Marine	TGE-Marine	TGE-Marine
Type and Construction	Independent, Type C, Cylindrical	Independent, Type C, Bi-Lobe	Independent, Type C, Cylindrical
Cargo Tank Number	2	2	2
Tank material	9% Ni Steel	9% Ni Steel	9% Ni Steel
Design temperature (°C)	-163	-163	-170
Design pressure (bar g)	4.0	4.5	3.5
Cargo density (kg/m^3)	500	500	500
LNG Bunker Transfer			
LNG Bunkering rate (m^3/h)	600	840	1,100
LNG Bunkering equipment	Flexible hose(s) Hose handling cranes: 2x telescopic type	Flexible hose(s) Hose handling cranes: 1x single boom type 1x knuckle boom type	1x FMC BOA transfer arm Hose handling crane: 1x single boom type (for elevated manifold/fenders)
Boil-Off Gas Handling			
Boil-Off as fuel	Y Main & Auxiliary Engines, Boiler	Y Main & Auxiliary Engines, Boiler	Y Main Diesel Generators, Boiler
Boil-Off burning	Y GCU	Y GCU	Y GCU
Re-liquefaction	N -	N -	Y 8.2 t/day
Compression	N -	N -	N -
Pressure build-up	Y IMO Type C tank	Y IMO Type C tank	Y IMO Type C tank
Propulsion System			
Propulsion arrangement	4 stroke DF direct drive	4 stroke DF direct drive	4 stroke DFDE
Main Engines – Type	2x Wartsila 9L20DF	1x Wartsila 6L34DF	3x Wartsila 8L20DF
Main Engines – Power	3,330 kW (2x 1,665 kW)	3000 kW	4,440 kW (3x 1,480 kW)
Propeller type	2x CPP	1x CPP	2x FPP
Rudder type	2x spade with flap	1x spade	2x spade with flap
Speed (knots)(Td, 85% power)	13.1	13.5	13.0
Tunnel Thruster(s) – Bow	2 x 270 kW	1x 600 kW (Retractable)	1x 650 kW
Tunnel Thruster(s) – Stern	-	1x 400 kW	-
Sources: Technical Data	[5] [6]	[7] [8]	[9] [10]

3.1 LBV STS LNG BUNKER TRANSFERS

At the time of writing LNG bunkering operations by Ship to Ship (STS) cargo transfer have been carried out by two of the purpose built LNG Bunkering Vessels (LBV) delivered in 2017.

These bunkering operations were carried out with the receiving ship and LNG Bunkering Vessel (LBV) double banked either alongside a jetty or in sheltered waters offshore. Dates of these first bunkering operations as well as the first receiving ship (Joint press release, 2017), (Tradewinds, 2017a) are indicated in Table 1.

Each of these 2017 deliveries of purpose designed LNG Bunkering Vessel (LBV) is being coupled with gas fuelled ships in formal agreements and contracts for the supply LNG bunker fuel. It is expected that this situation may prevail pending wider adoption of LNG fuel, increased LNG bunker fuel supply, and availability of alternative bunkering facilities.

3.2 FUTURE LBV DESIGNS AND PROJECTS

At the time of writing a 2200m^3 capacity LNG Bunkering Barge (LBB) *Clean Jacksonville* with a membrane (GTT Mk III type) cargo containment system (CCS) and LNG bunker loading mast (GTT Reach4® type) is being built at Conrad Shipyard for entry into service in 2018 for Clean Marine Energy. This LNG Bunkering Barge (LBB) is planned to be operated from Jacksonville on charter to TOTE Maritime.

An additional LNG Bunkering Vessel (LBV) will also commence construction during November 2017 for delivery in second half of 2018. This 2018 delivery is shown in Table 5 in a comparison of particulars with two recent design developments, with Lloyd's Register approval, of LBV. Artist impressions/renderings of these future LBVs are shown in Figure 3.

In addition there are two further contracts for newbuilding LNG Bunkering Vessels (LBV) to be delivered in 2019 for which particulars are unavailable for publication at the time of paper writing;

- Two 7500m^3 LBV contracted in May 2017 by Stolt Nielsen Gas with Keppel Singapore
- One 7500m^3 LBV, with membrane type cargo containment system, contracted by KOGAS with Samsung Heavy Industries

Top: *Karios*, a 7500m^3 capacity LBV design, with Babcock LGE Process cargo tank and LNG system, to commence construction in November 2017 at Hyundai Mipo Dockyard *(courtesy Babcock-LGE Process)*

Middle: SENER 8000m^3 capacity LBV design *(courtesy SENER)*

Bottom: HMD 6600m^3 capacity LBV design *(courtesy Hyundai Mipo Dockyard)*

Figure 3: Artist impressions/renderings of future LBVs

Table 2: Comparison of particulars of future LNG Bunkering Vessel (LBV) projects

LNG Bunkering Vessel	***Karios***	-	-
Cargo capacity (m^3)	7500	8000	6600
Designer	Hyundai Mipo Dockyard, Ulsan, Republic of Korea	SENER, Madrid, Spain	Hyundai Mipo Dockyard, Ulsan, Republic of Korea
Shipbuilder		-	
Yard/NB/Hull No.	8250	-	-
Classification	Lloyd's Register	Lloyd's Register approved	Lloyd's Register approved
LNG Bunkering Operation			
LNG Terminal	Klaipeda	-	-
Owner	Babcock Schulte Energy	-	-
Charter Party	Nauticor	-	-
Planned delivery date	09-2018	-	-
Main Dimensions			
Length over all, LOA (m)	About 117	114.8	About 119
Length btw. perp., LBP (m)	113.5	112.0	116.0
Breadth moulded, B (m)	20.0	20.4	19.4
Depth moulded, D (m)	10.3	8.5	10.6
Design draught, Td (m)	5.2	5.5	6.0
Deadweight and Tonnage			
Deadweight (tonnes) (Td)	4100	4315	4900
LNG Cargo Tanks			
Cargo tank & system design	Babcock LGE Process	SENER	Babcock LGE Process/ Hyundai Mipo Dockyard
Type and Construction	Independent, Type C, Cylindrical	Independent, Type C, Cylindrical	Independent, Type C, Cylindrical
Cargo Tank Number	2	2	2
Tank material	9% Ni Steel	9% Ni Steel	9% Ni Steel
Design temperature (°C)	-165	-163	-163
Design pressure (bar g)	3.75	4.5	4.0
Cargo density (kg/m^3)	500	500	500
LNG Bunker Transfer			
LNG Bunkering rate (m^3/h)	1,150	1,600	1,100
LNG Bunkering equipment	Flexible hose(s)	Flexible hose(s)	Flexible hose(s) OR Transfer arm
	Hose handling cranes: 2x knuckle boom type	Hose handling crane: 1x single boom type	Hose handling crane: 1x single boom type
Boil-Off Gas Handling			
Boil-Off as fuel	Y – Main Diesel Generators	Y – Main Diesel Generators, Boiler	Y – Main & Auxiliary Engines, Boiler
Boil-Off burning	N – -	N – -	Y – GCU spec. option
Re-liquefaction	N – -	N – -	Y – Spec. option
Compression	Y – 220 bar CNG	N – -	N – -
Pressure build-up	Y – IMO Type C tank	Y – IMO Type C tank (35 day holding time)	Y – IMO Type C tank
Propulsion System			
Propulsion arrangement	4 stroke DFDE	4 stroke DFDE	4 stroke DF direct drive
Main Engines – Type	3x Wartsila 8L20DF	-	1x Hyundai 6H35DFP
Main Engines – Power	4,800 kW (3x 1,600 kW)	6,000 kW (2x 3,000 kW)	3,000 kW
Propeller type	2x Azimuth thrusters (2x 1,900 kW)	2x Azimuth thrusters	1x CPP
Rudder type			1x spade with flap
Speed (knots)(Td, 85% power)	12.5	15.0	13.0
Tunnel Thruster(s) – Bow	2x 320kW	1	1
Tunnel Thruster(s) – Stern	-	-	1
Sources: Technical Data	[13] [14]	[15]	[16]

4. LBV SPECIFICATION AND DESIGN

Some of the basic considerations for specification, and design, of an LNG Bunkering Vessel (LBV) are described in the following section.

4.1 LBV ARRANGEMENT AND LAYOUT

LBV designs with cylindrical IMO Type C tanks are typically arranged with a continuous strength deck with poop and forecastle. Upper domes of cylindrical IMO Type C cargo tanks protrude through the continuous strength deck in way of each cargo hold. A weathertight partial, or continuous, cover of exposed upper tank dome is fitted. The dome cover can be a continuous construction with integration structurally into a raised poop and forecastle. Dome cover mounted platforms are arranged for cargo manifold, cargo handling equipment and cargo processing rooms.

LBV designs with bi-lobe IMO Type C tanks, prismatic IMO Type B tanks and membrane cargo containment system may be arranged with a continuous strength deck extending over the whole length of ship. Alternatively a continuous trunk deck may be fitted structurally integrated into a raised poop and forecastle.

Cargo holds are subdivided by transverse bulkhead(s) and partial double skinning may be adopted in way of the cargo holds for ballast water carriage.

Accommodation and engine room are typically arranged aft. In one case, where adoption of a Dual Fuel Diesel Electric (DFDE) propulsion system allowed, the accommodation and engine room has been arranged forward. See Figure 3 top image of 7500m^3 capacity LBV Karios.

For this case the forward accommodation and engine room layout allows for control of trim without the use of sea ballast water. As designed the ship is to be capable of adjustment of trim (and list) by using three pairs of fresh water tanks.

4.2 LNG BUNKER CARGO TANK TYPES

Considerations for LNG bunker cargo tanks for an LNG Bunkering Vessel (LBV) include:

- Flexibility in partial cargo loading
- Ease of Boil-Off Gas (BOG) management
- Low Boil-Off Rate (BOR)
- Low cost, low weight, maximized cargo volume

LNG Bunkering Vessels (LBV) that entered service in 2017 adopted either Cylindrical or Bi-lobe IMO Type C tank type, see Figure 4.

For LBV applications a membrane cargo containment system, with prismatic tank configuration, is a proven LNG tank type alternative to an IMO Type C tank.

The main technical advantage of an IMO Type C tank is the ease of pressure build up or accumulation as a means of LNG bunker transfer and BOG treatment.

The main technical advantage of membrane system is the possibility of high hull volume utilization and space efficiency. On the other hand the main technical drawback of an IMO Type C cylindrical tank is a low hull volume utilization and space efficiency when compared with IMO Type C Bi-Lobe tank and membrane systems applied to prismatic tanks.

Figure 4: IMO Type C tanks

Top: Cylindrical tank *(Courtesy of STX O & S)*
Bottom: Bi-lobe tank *(Courtesy of Royal Bodewes)*

4.3 LNG BUNKER CARGO TANK CAPACITY

A basic consideration for specification of LNG bunker cargo capacity for an LBV is it's compatibility with the bunker fuel supply requirement of receiving ships.

In the following examples LNG bunker fuel capacities of large commercial ships contracted for construction, or in service, utilizing gas as a fuel for main propulsion are described:

- **Two (2) Handysize Bulk Carriers** (ESL Shipping website, 2016): to be delivered to ship owner ESL Shipping in 2018 each with a single cylindrical IMO Type C LNG bunker tank of 400m^3 capacity arranged at main deck level aft of the accommodation. LNG bunker fuel is to supplied by agreement between ESL Shipping and SkanGas from an LNG terminal in the Port of Tornio, Finland (SkanGas, 2017)
- **Four (4) Aframax Oil Tankers** (Tradewinds, 2017b) to be delivered to ship owner Sovcomflot in 2018 and 2019 to be fitted with two cylindrical IMO Type C LNG bunker tanks, each of 850m^3 bunker capacity, located on the upper deck forward of the accommodation.
- **Two (2), 3100 teu, Container Ships** (GD NASSCO, 2016): delivered to ship owner Tote Maritime in October 2015 and February 2016 for trade from US Gulf Coast to Puerto Rico in US (SECA) waters. Two cylindrical IMO Type C LNG bunker tanks are arranged at main deck level aft (of a ¾ aft accommodation block) with bunker fuel capacity of approximately 1,800m^3. These ships are to be supplied with LNG bunker from a 2200m^3 LNG bunkering barge (LBB) *Clean Jacksonville* operated on charter to TOTE Maritime (TOTE Maritime, 2017)
- **Two (2), 4000 unit, Pure Car and Truck Carriers (PCTC)** (TGE Marine Gas Engineering, 2016): delivered in 2016 to ship owner UECC for North Sea and Baltic Sea trade. A single cylindrical IMO Type C LNG bunker tank of 800m^3 bunker capacity is arranged below the main deck.
- **Two (2) 4500 unit Pure Car and Truck Carriers (PCTC)** (Tradewinds, 2016): to be delivered in 2019 to ship owner SIEM Car Carriers. Cylindrical IMO Type C LNG bunker tank, or tanks, with 3000m^3 bunker capacity to be arranged below the main deck.

LNG bunker cargo capacity of existing, and future, LNG Bunkering Vessels (LBV) described in this paper, which fall within a range 5000m^3 to 8000 m^3, and appear well matched to expected LNG bunker tank capacities of example receiving ships considered above.

4.4 PROPULSION ARRANGEMENT

Main requirements for propulsion system selection and specification for an LNG Bunkering Vessel (LBV) are:

- Utilizing LNG (BOG and Vaporized LNG Cargo)
- Redundancy
- Low fuel oil/gas consumption
- High performance in manoeuvrability
- Reasonable initial cost

A single direct drive Dual Fuel (DF) 4 stroke engine together with reduction gear and either fixed pitch (FPP) or controllable pitch (CPP) propeller is considered the most economical propulsion system for an LBV. To improve the manoeuvrability, at least one transverse thruster is ordinarily specified.

Alternatives, and additional, specification options for the propulsion arrangement have been adopted for LNG Bunkering Vessels entering into service in 2017 and 2018 as shown in the tabular comparisons in Tables 1 and 2. These propulsion arrangement options have included:

- Twin propellers (FPP or CPP), see Figure 5
- Single/twin rudders with flaps, see Figure 5
- Additional transverse tunnel thruster
- Dual Fuel Diesel Electric (DFDE) system
- Twin azimuthing thrusters (with DFDE)

Figure 5: Twin FPP with twin rudders with flaps *(Courtesy of STX O & S)*

5. LNG BUNKER TRANSFERS

Some of the equipment specification and technical considerations associated with Ship to Ship (STS) LNG bunker transfers are described in the following section.

5.1 EQUIPMENT FOR LNG TRANSFER

Consideration of on-board equipment for LNG transfers is often considered a binary choice between:

- Loading arm installation, OR
- Use of (cryogenic) flexible hoses

However for a Ship to Ship (STS) LNG bunker transfer there are complex considerations for LNG transfer equipment, as well as many available specification options and alternatives. Two examples are described below to offer an insight into possible specification considerations for LNG transfer and bunkering interface equipment.

First example is the Reach4® LNG bunker loading mast shown in Figure 6 (LNG Industry Magazine, 2015). Offered by GTT this is a rotating and luffing mast of around 20 m boom length.

Figure 6: Reach4® LNG bunker loading mast
(Courtesy of GTT)

A stated key advantage of Reach4®: a simple emergency disconnect system with breakaway couplings. The first installation of a Reach4® LNG loading mast is for a 2200m^3 capacity LNG Bunkering Barge (LBB) *Clean Jacksonville* to be delivered from Conrad Shipyard to Clean Marine Energy.

Second example is a "hybrid" LNG bunker transfer system; combining features of both a loading arm as well as a flexible hose based transfer.

Developed by collaboration between KLAW LNG and Houlder Ltd the "KHobra" system is arranged to be deployed using a standard knuckle boom crane (Houlder/Klaw, 2017). "KHobra" utilizes a customised frame mated to LNG bunker loading and vapour return hoses as shown in Figure 7.

A stated key advantage of "KHobra": all connections are accommodated in a single lift with very significant reductions in connection times.

A set of pneumatic fenders, see Figure 8 example, are typically provided as part of LBV bunker transfer equipment for double bank mooring for Ship to Ship (STS) LNG transfer.

Figure 7: "KHobra" hybrid bunker transfer system including lower image close up of customised frame *(Courtesy of KLAW LNG and Houlder Ltd)*

Figure 8: Pneumatic fenders on LBV *Coralius (Courtesy of Royal Bodewes)*

For controlled rapid shut down of LNG transfer operations during an emergency the LNG bunkering system (as well as the LNG fuel system on-board the receiving vessel) is specified with Emergency Shut-Down (ESD) valves. LNG Bunkering Vessel (LBV) shall also be arranged to accommodate different link Emergency Shut-Down (ESD) systems (e.g. Universal Safety Link (USL) system) available.

5.2 MEANS OF LNG TRANSFER

LNG bunker cargo tanks are usually provided with deep well submerged pumps. In accordance with IGC Code, when cargo transfer is by means of non-accessible (e.g. submerged) cargo pumps, two separate means of LNG transfer shall be provided.

The IGC Code accepts the possibility to carry out transfer of the cargo by gas pressurization. Where IMO Type C cargo tanks are selected for an LNG Bunkering Vessel consideration could be given to the adoption of a primary means of discharge by mechanical pumps and a secondary means via gas pressurization.

To maximize the amount of energy transferred it is desirable for LNG bunker cargo to be carried at its design temperature at atmospheric (1 bar g) pressure. Atmospheric conditions of carriage can also be achieved for IMO Type C cargo tanks, as well as, membrane type cargo tanks and cargo carriage of LNG bunker cargo at atmospheric pressure in an IMO Type C cargo tank allows for pressure accumulation (as one means of cargo transfer) during bunkering operations.

Higher LNG bunker cargo transfer rates generate a higher volume of Boil-Off Gas (BOG) so one possible limitation of LNG Bunker Cargo transfer rate is the capacity of the LNG Bunkering Vessel (LBV), as well the LNG bunker receiving vessel, to handle the Boil-Off Gas (BOG).

5.3 MEASUREMENT OF LNG TRANSFER

Custody Transfer System (CTS) is typically in use by LNG carriers to indicate the LNG transfer volumes using level gauges only. Third party laboratory analysis of LNG composition is necessary in a Custody Transfer System (CTS) to measure cargo delivery. For LNG bunker transfers Custody Transfer System (CTS) is not able to measure the exact quantity of LNG transfer from LBV to receiving ship alone as the composition and properties of LNG at the receiving ship will be slightly different from the LBV. For this reason, for LNG bunkering operations, the use of mass flow meters in combination with calorific value conversion (using a gas chromatograph) into an energy transfer is the preferred option for custody transfer of LNG bunkers.

The Society for Gas as a Marine Fuel (SGMF) has published a detailed Guidance for Quality and Quantity of the LNG transfer during bunkering operations (Society for Gas as a Marine Fuel, 2015).

5.4 BOIL-OFF GAS (BOG) HANDLING

Depending on the tank size and type of insulation LNG Boil-Off Gas (BOG) may be generated at Boil-Off Rates (BOR) of 0.2% to 0.6% (BOR) of tank volume per day for LNG Bunkering Vessels (LBV). Depending on LNG bunker tank arrangements on board the receiving ship may also generate LNG Boil-Off Gas (BOG) at similar Boil-Off Rates (BOR).

However it is during LNG bunker transfer operations when significant additional Boil-Off Gas (BOG) is generated. This generated Boil-Off Gas (BOG) requires safe handling during the LNG transfer operations without any emissions, where emission control regulations typically apply, in prevailing bunkering locations.

The LNG receiving ship may have its own means to handle the Boil-Off Gas (BOG) generated within LNG bunker fuel tanks. However, it is also typically expected that the LNG Bunkering Vessel (LBV) will assist, partially or completely, the receiving vessel to handle the additional Boil-Off Gas (BOG) generated as a result of LNG bunker transfer operation.

For an LNG Bunkering Vessel (LBV) a combination of following means will be necessary for Boil-Off Gas (BOG) handling:

- Use of Boil-Off Gas (BOG) as a fuel: for power generation and/ or propulsion
- Re-liquefaction of Boil-Off Gas (BOG)

- Storage by compression of Boil-Off Gas (BOG): in one LBV case, *Karios*, BOG is compressed and stored in separate CNG tanks for later use as a fuel
- Pressure build-up or accumulation: which is feasible in case of IMO Type C LNG tank types
- In a final contingency, or in an emergency, burning of Boil-Off Gas (BOG) by the LBV in a Gas Combustion Unit (GCU) (where fitted)

A disadvantage of pressure accumulation is that it will cause LNG temperature rise and additional Boil-Off Gas (BOG) generation. LNG composition changes may result which could affect methane number, quality of LNG to be delivered, and LNG temperature, which in turn may lead to higher Boil-Off Gas (BOG) generation during LNG bunker transfer. Use of Boil-Off Gas (BOG) as a fuel is then the desired, and most economical, approach for Boil-Off Gas Handling, especially as LBV is expected to be operating in an emission and environmentally controlled area.

6. FUTURE LBV SPECIFICATIONS

Future LNG Bunkering Vessel (LBV) specifications and designs will be influenced by developments in LNG technology, as well as, an expected increased demand for LNG bunker fuel as more commercial ships adopt gas as a fuel. Described below, as considerations for the specification of a future LNG Bunkering Vessel (LBV), are two possible influences on future LBV designs:

- Introduction of new LNG tank types and designs
- Adoption of much larger LNG bunker tanks for very large commercial ships

6.1 NEW LNG TANK TYPES AND DESIGNS

There are a number of new tank types and designs being considered which may be adopted in future LNG Bunkering Vessels (LBV) these include:

- Tri-lobe and multi-lobe IMO Type C tank types
- New lattice technology tank types
- New design variants of membrane type cargo containment systems

The principal driver for adoption of new tank types is to improve hull volume utilization and space efficiency. New design variants of membrane cargo containment systems are to facilitate pressure accumulation as a means of LNG bunker transfer, as well as Boil-Off Gas (BOG) treatment, with an increase in LNG bunker fuel tank pressure above 0.7 bar g.

6.2 LARGER LNG BUNKER TANKS

Ultra Large Container Ships (ULCS) of 20,000 teu cargo capacity may consume 500m^3 of LNG per day. If endurance for a round-trip on an Asia-Europe service is required this would correspond to 40 days operations with an LNG bunker capacity of about 22,000m^3.

In addition there are other large commercial ships, such as Ore Carriers of 350,000 to 400,000 dwt tonne size for trade from Brazil to China, where hull space has been allocated for future installation of large LNG bunker tanks >25,000m^3 capacity for round-trip endurance (on conversion to gas as a fuel operation).

If Ship to Ship (STS) LNG bunker transfer is anticipated with implementation of gas as a fuel on Ultra Large Container Ships (ULCS), where short port turnaround times prevail, this may catalyse a future LNG Bunkering Vessel (LBV) design with;

- An increased LNG bunker cargo capacity, possibly up to 30,000m^3 in size
- An increase in LNG bunker cargo transfer rate in proportion to the increase in volume of LNG bunker to be transferred to the receiving ship
- Faster connection and disconnection speed for LNG bunkering interfaces encouraging the adoption, and installation, of specialised and purpose designed LNG transfer equipment similar to solutions described earlier

7. CONCLUSIONS

A new type of purpose designed LNG Bunkering Vessel (LBV) has emerged to meet the need for LNG bunkering operations: a combination of a small scale LNG carrier with specialist equipment for handling and Ship to Ship (STS) LNG bunker cargo transfer.

Three purpose built LNG Bunkering Vessels (LBV) have entered service in North European waters in 2017 with initial deployment in the port of Antwerp, port of Rotterdam, and for harbour and coastal service off Scandinavia and in the Baltic Sea.

Each of these 2017 deliveries of purpose designed LNG Bunkering Vessel (LBV) are being coupled with gas fuelled ships in formal agreements and contracts for the supply of LNG bunker fuel. LNG bunker cargo tank capacities of LBVs entering into service in 2017, 2018 and 2019 appear to be well matched to the LNG bunker fuel tank requirements of gas fuelled ships in service and being built for entry into service in 2018 and 2019.

Ultra Large Container Ships (ULCS) of 20,000 teu cargo capacity may consume 500m^3 of LNG per day. If Ship to Ship (STS) LNG bunker transfer by LBV is anticipated for Ultra Large Container Ships (ULCS) this may encourage the development of LNG Bunkering Vessel (LBV) designs with larger LNG cargo capacity, faster bunker transfer rates and faster connection and disconnection of bunkering interfaces.

8. ACKNOWLEDGEMENTS

The authors wish to thank Lloyd's Register EMEA and Asia for supporting paper publication. The authors wish to also thank industry contributors for their permissions and consent to use credited photographs and reference documents in support of paper publication. Any views expressed are from the authors alone and do not necessarily represent those of Lloyd's Register.

9. REFERENCES

1. IHS FAIRPLAY & CLARKSONS SHIP RESEARCH COMBINED DATA SOURCES, 'Global Existing Fleet and Order-Book of LNG-Fuel Enabled Vessels', Lloyd's Register, 2017
2. SOCIETY FOR GAS AS A MARINE FUEL, 2017
3. BENITO, L., 'LNG Bunkering Infrastructure Survey: The outlook of Ports on provision of LNG Bunkering facilities', Lloyd's Register, 2014
4. DANISH MARITIME AUTHORITY, 'North European LNG Infrastructure Project: A feasibility study for an LNG filling station infrastructure', Danish Maritime Authority, 2012
5. ROYAL INSTITUTION OF NAVAL ARCHITECTS, 'Significant Ships of 2016', RINA, 2017
6. TGE MARINE GAS ENGINEERING DATASHEET 2735, '5,100 cbm LNG Bunker-Vessel "Engie Zeebrugge" for Gas4Sea (Engie, NYK, Mitsubishi)'', TGE Marine Gas Engineering GmbH, 2017a
7. ROYAL BODEWES WEBSITE DATASHEET, '5800 m3 LNG carrier', Royal Bodewes, 2017
8. TGE MARINE GAS ENGINEERING DATASHEET 2826, '5,800 cbm LNG Bunker Vessel "Coralius" for Siruis Veder Gas AB', TGE Marine Gas Engineering GmbH, 2017b
9. STX OFFSHORE & SHIPBUILDING POCKET PLAN, 'Shell 6.5 K LNG Bunkering Vessel', STX O & S, 2014
10. TGE MARINE GAS ENGINEERING DATASHEET 2960, '6,500 cbm LNG Bunker-Vessel "Cardissa" for Shell', TGE Marine Gas Engineering GmbH, 2017c
11. JOINT PRESS RELEASE (ENGIE, MITSUBISHI CORPORATION, NYK, FLUXYS, PORT OF ZEEBRUGGE, UECC), 'Ship-to-ship LNG bunkering service started in the port of Zeebrugge', Port of Zeebrugge, 14th June 2017
12. TRADEWINDS NEWS ARTICLE, 'SkanGas marks first Ship-to-Ship LNG bunkering', Tradewinds, 19th September 2017a
13. HYUNDAI MIPO DOCKYARD POCKET PLAN, '7,500 m^3 Class LNG bunkering vessel', Hyundai Mipo Dockyard, 2017a
14. LNG SHIPPING 2017 PRESENTATION, 'LNG gas supply vessels – setting the standard', Babcock LGE Process, June 2017
15. SENER PRESENTATION TO LLOYD'S REGISTER, 'LNG Bunkering Vessel', SENER, September 2016
16. HYUNDAI MIPO DOCKYARD POCKET PLAN, '6,600 m^3 Class LNG bunkering vessel', Hyundai Mipo Dockyard, 2017b
17. ESL SHIPPING WEBSITE, 'Press Room Blog', ESLShipping, 29th July 2016
18. SKANGAS PRESS RELEASE, 'LNG from SkanGas to ESL Shipping's bulk carriers', SkanGas, 24th August 2017
19. TRADEWINDS NEWS ARTICLE, 'SCF and Shell make pioneering move in LNG-fuelled tankers', Tradewinds, 16th March 2017
20. GENERAL DYNAMICS NASSCO PRESS RELEASE, 'General Dynamics NASSCO Delivers Second Liquefied Natural Gas Powered Containership', General Dynamics NASSCO, 22nd January 2016
21. TOTE MARITIME PRESS RELEASE, 'JAX LNG, TOTE Maritime's LNG supplier, has received approval for their LNG facility and barge-to-ship bunkering', TOTE Maritime, 22nd August 2017

22 TGE MARINE GAS ENGINEERING PRESS RELEASE, 'World's first Pure Car/Truck Carrier with high pressure LNG fuel gas system put into operation', TGE Marine Gas Engineering GmbH, 23rd September 2016

23 TRADEWINDS NEWS ARTICLE, 'Volkswagen to charter Siem Car Carrier LNG PCTCs', Tradewinds, 6th October 2016

24. THE SOCIETY FOR GAS AS A MARINE FUEL, 'Contractual Guidelines Quantity and Quality', Society for Gas as a Marine Fuel, September 2015
25. LNG INDUSTRY MAGAZINE ARTICLE, 'Unlocking the LNG Fuel Supply Chain', LNG Industry magazine, October 2015
26. HOULDER/KLAW LNG PRESENTATION, 'The Houlder-KLAW "Khobra" Transfer System', Houlder Ltd, April 2017

10. AUTHORS BIOGRAPHY

Woo-Jeong Choi holds the current position of Lead Sales/Client Relationship Manager at Lloyd's Register Asia. Prior to joining Lloyd's Register he has had 24 years of shipyard experience includes sales engineering, shipbuilding control, commissioning, trials and project management involvement in projects for the build of LNGC and LBV.

Jose Navarro holds the current position of Principal Gas Technology Specialist at Lloyd's Register EMEA. He provides expert advice to Clients globally on gas technology and contributing to the creation of specialist rule and regulatory requirements. He has over 20 years' experience of gas technology and newbuilding in Korea, Japan and China.

Robert Tustin holds the current position of Technical Consultant, New Construction for Lloyd's Register EMEA. He is responsible for delivery of pre-tender and pre-contract consultancy advice for non-standard and special ship types. He has 21 years' experience with ship new construction based in North Asia.

INNOVATIVE TANK FOR SMALL SCALE LNG BUNKER VESSELS, COASTAL VESSELS AND FUEL TANKS

T Lamb, University of Michigan, USA, **R Ramoo**, Altair ProductDesign, Inc., USA, **B W Eng**, Altair Taiwan, Inc., **S Egambaram**, **M N Nadzri**, Altair Malaysia Sdn. Bhd.

SUMMARY

Small Scale LNG has become inclusive of all aspects of LNG storage and transport in recent years. Many ship owners are entering the LNG bunker vessel market. To date they have all used cylindrical or bi-lobe tanks. Designs have also been prepared using a 2 Bar variant of the membrane system and a recent order using the Korean KC-1 membrane system. An approach to satisfying the complete logistical needs is presented utilizing the SEMI-CDTS for storage on land, on barge and in LNG Bunker Vessels. The paper describes how a new tank design provides a superior alternative to the cylinder and bi-lobe designs, and shows its flexibility to fit new and converted bunker and small coastal vessels as well as fuel tanks in several types of ships, including cruise ships.

NOMENCLATURE

BoG	Boil Off Gas
Btu	British Thermal Unit
CM	Cubic Meters
CDTS	Cubic Donut Tank System
LNG	Liquid Natural Gas
m	meters
MDO	Marine Diesel Oil
MHO	Marine Heavy Oil
OSV	Offshore Supply Vessel
SHP	Shaft Horsepower KW
t	metric ton

1. INTRODUCTION

Small Scale LNG has become inclusive of all aspects of LNG storage and transport in recent years. The increasing use of LNG as a ship propulsion fuel requires small LNG vessels to bunker the LNG fuelled ships. Many ship owners are getting into the LNG bunker vessel market. To date they have all used cylindrical or bi-lobe tanks. Designs have also been prepared using a 2 Bar variant of the membrane system and a recent order using the Korean KC-1 membrane system. One of the authors converted offshore supply vessels into crab fishing vessels for the Alaskan Crab industry. He agrees that cost effective LNG bunker vessels could be quickly obtained by converting OSVs based on the cost and design simplicity of the conversion.

The paper presents an approach to satisfying the complete logistical needs, excluding from gas field to initial storage which would be the same for all options, utilizing the SEMI-CDTS for storage on land, on barge and in LNG Bunker Vessels. The paper describes how a new tank design provides a superior alternative to the cylinder and bi-lobe designs, and shows its flexibility to fit new and converted bunker and small coastal vessels as well as fuel tanks in several types of ships including cruise ships. A description of the development of the new tank, including its structural analysis, is presented.

While large liquefied natural gas carriers have been around for some time, carrying large cargoes of LNG from source to user, where the BoG has been used in boilers to generate steam for steam turbine main engines, it is only recently that the focus has changed to using LNG as a fuel for both LNG Carriers and non-LNG carrying ships. Tankers, ferries and work boats have already been built or converted to use LNG as their fuel. To date the LNG containment tanks have been cylindrical in shape which suffers from low volumetric efficiency thus taking up a lot of space on the ship. ALTAIR's SEMI-CDTS offers a solution that has better volumetric efficiency than the cylindrical or bi-lobe tanks, lower weight, reduced vertical center of gravity. and lower installed cost than the others.

2. HISTORY

The Cubic Donut Tank System (CDTS) was first developed and patented by Thomas Lamb in 1976. The objective being that it is as efficient as the spherical tank system for hydrostatic containment and has the volumetric efficiency of membrane tanks for large LNG Carriers. From 2005, Altair's Award-winning design synthesis technology Optistruct® has been used to improve on the unique geometry of the CDTS to provide the best trade-off between spatial & structural efficiencies. Figure 1 shows the development for the CDTS and for the SEMI-CDTS specifically for small LNG Bunker Vessels and Fuel Tanks. Figure 2 shows the application of the CDTS family of tanks.

3. DESCRIPTION

The SEMI-CDTS tank is a derivative of the CDTS tank. The CDTS tank consists of twelve intersecting cylinders forming the twelve edges of a cube and closure surfaces on all faces of the cube. It has been described in detail in a number of papers (Lamb & Ramoo, 2009a, 2009b, 2010 & 2016). The development of the SEMI-CDTS from the original CDTS was achieved by eliminating the upper four cylinders and the vertical cylinders connecting those to the lower four cylinders. Thus, the SEMI-CDTS consists of four intersecting cylinders with, or without closure surfaces for smaller volumes or high-pressure applications, on the top and bottom faces.

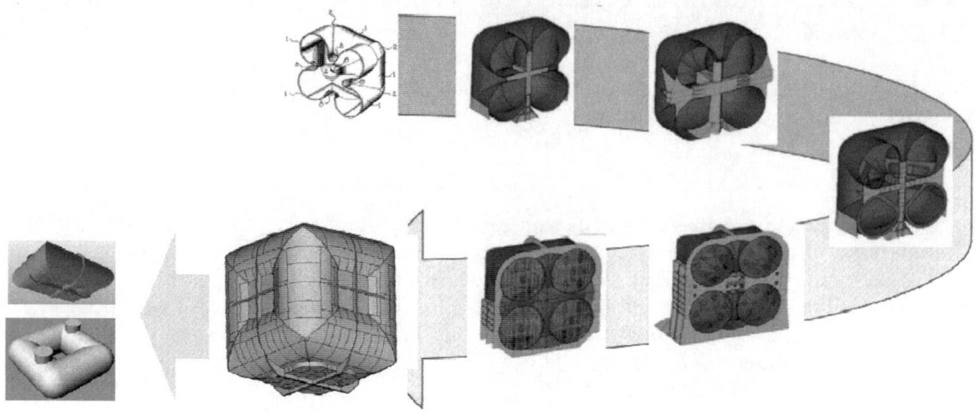

Figure 1. CDTS Development

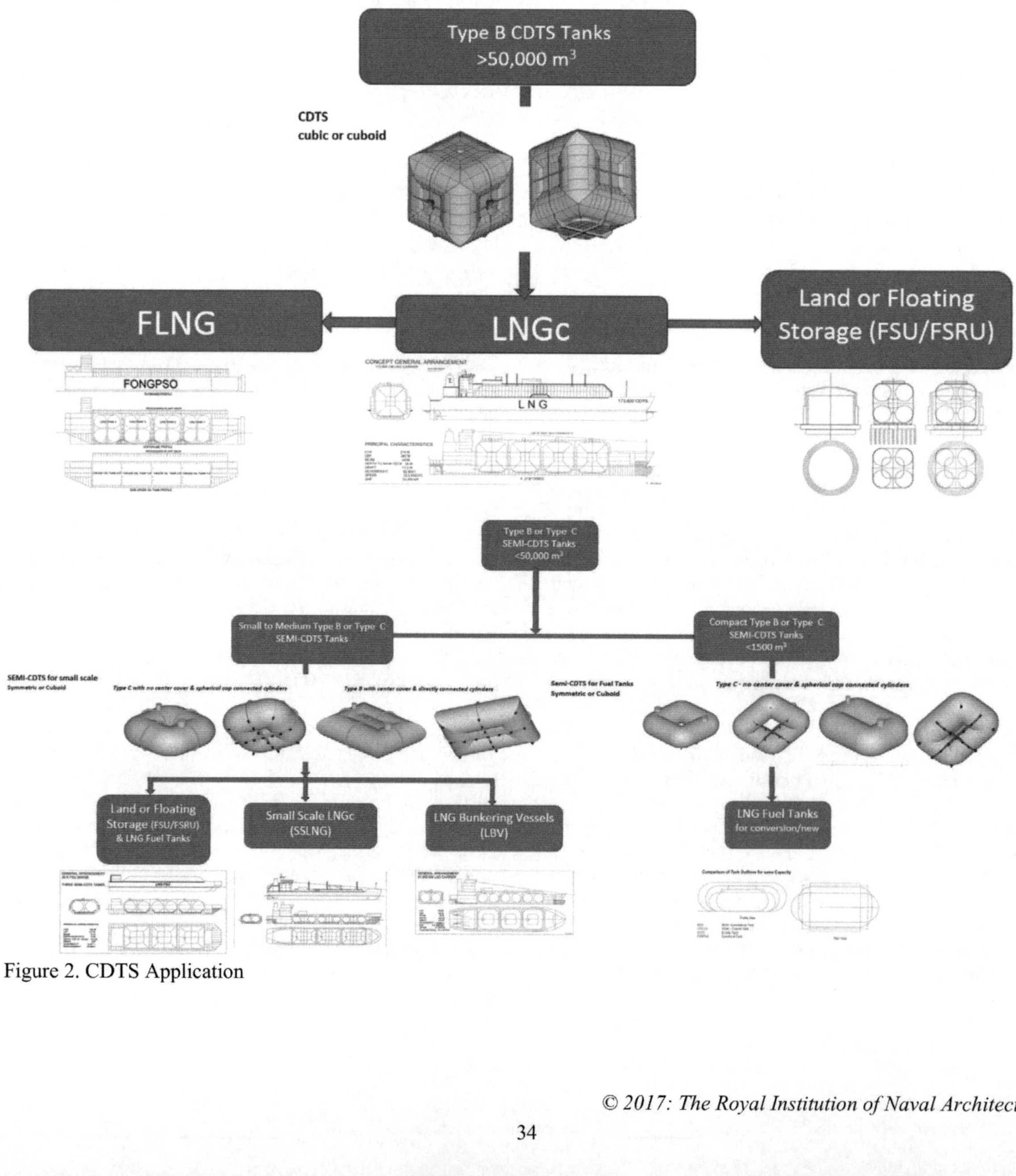

Figure 2. CDTS Application

The SEMI-CDTS tank can be SYMMETRICAL (same length and width) or CUBOID (different length and width). The tank is self-standing (independent) Type B or Type C and has no internal stiffeners, but there are Ring Webs in the cylinders and corners. For the Type B Tank, the cylindrical segments can be directly interconnected whereas a spherical cap is required for the Type C variant. The tanks are insulated with a PUF foam to reduce boil-off.

4. TANK BENEFITS/COMPARISON

The SEMI-CDTS has all the advantages of other self-standing independent tanks such as:

- completed tank inserted into ship
- minimum work on ship
- no restriction on partial filling of tanks due to sloshing loads
- construction schedule separation of tank from ship
- outsourcing of tank construction
- more cost effective than inserting a new midbody or building a new stern
- reduced new building or conversion schedule

In addition, the use of the SEMI-CDTS results in the following advantages for small scale LNG carriers and platforms:

- reduction in ship or platform length for equivalent LNG capacity; 8% compared to Bi-lobe, 4% to Cylindrical and 2 % to membrane, or more capacity for same length
- reduced (estimated) cost of LNG carrier by 21% compared to Bi-lobe, 11% to Cylindrical and 15% to Membrane
- weight reduction; up to 12%/15% compared to cylindrical/bi-lobe tanks
- reduction in tank height, up to 28%/17% compared to cylindrical/bi-lobe tanks
- lower vertical center of gravity
- superior structural efficiency
- simple support system
- excellent BoG performance – estimated at ~0.37% with 300 mm PUF and ~0.22% with 500 mm PUF
- self-centering support system allows for thermal contractions while constraining tank against vertical, lateral and longitudinal motions

All combining to offer a cost-effective alternative to the other available and proposed tank systems. Figure 3 compares the outlines of the different tanks for an equal capacity. The SEMI-CDTS Symmetrical tank is shorter than the Cylindrical and Bi-lobe tanks. Both the symmetric and cuboid variants of the SEMI-CDTS tanks are wider than the Cylindrical and Bi-lobe tanks, but the benefit is in the tank height which are the same as the tank cylinder dimension. The smaller diameter cylindrical segments of the SEMI-CDTS tanks results in lower membrane stresses. The Cuboid variant allows for the specific design to suit the available space on ship rather than having to change the space to suit tank.

Figure 4 shows plots of tank capacity for different cylinder diameters for the CDTS Tank family and the Cylindrical and Bi-lobe Tanks. It also incorporates a table that shows the typical geometric proportions of the different tank types. The equivalent diameters for a given capacity can be read off the vertical axis. For example, a 100 CM tank would have diameters of 2.6 m for a Cuboid SEMI-CDTS, 2.9 m for a Symmetrical SEMI_CDTS, 3.25 m for the Bi-lobe and 3.7 m for the Cylindrical tank. Table I presents a comparison of the dimensions of the different tank types for a range of ship types. The height and length benefits of the CDTS and SEMI-CDTS are clear.

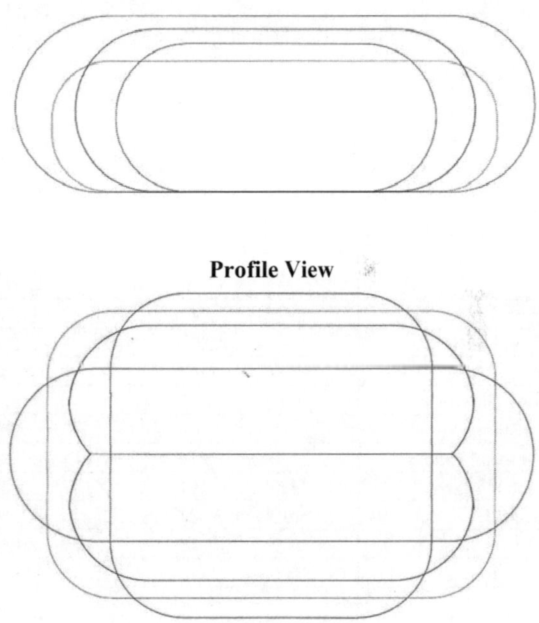

RED	SEMI- Symmetrical Tank
GREEN	SEMI – Cuboid Tank
BLUE	Bi-lobe Tank
PURPLE	Cylindrical Tank

Figure 3. Comparison of Tank Outlines for Same Capacity

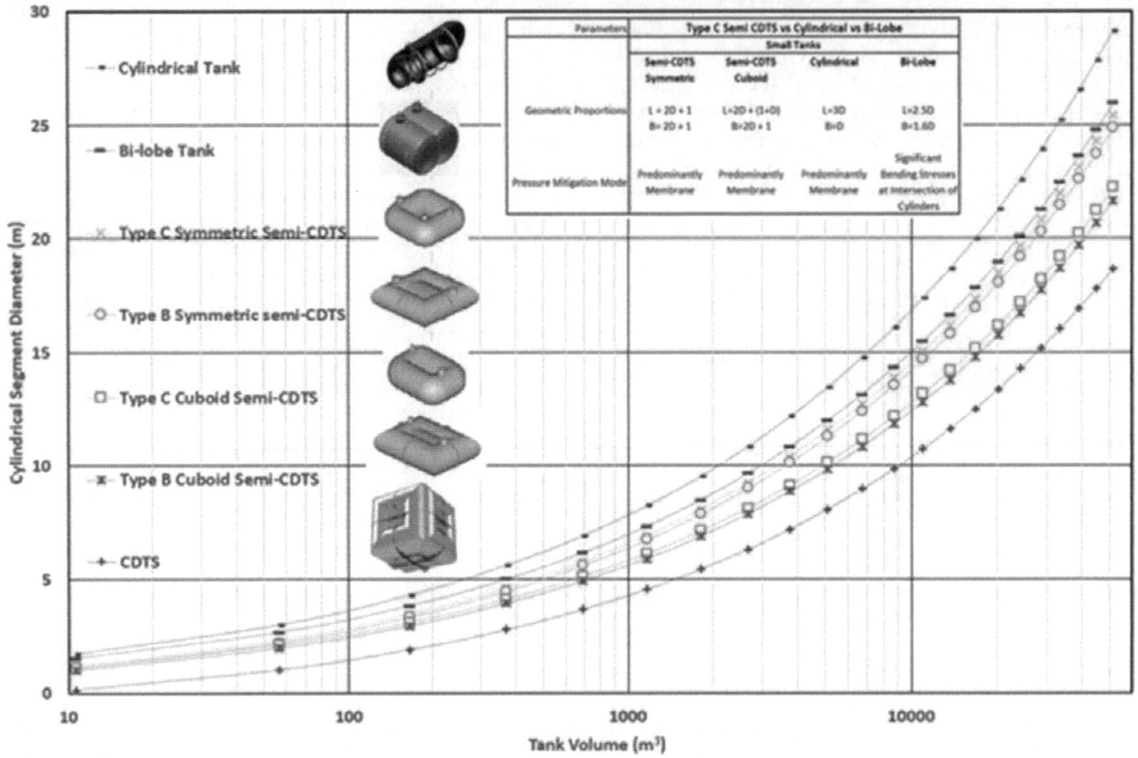

Figure 4. Tank Capacity/Diameter comparison for different Tank Type

Table I. Comparison of different fuel tanks for a range of vessels

	MDF CM	LNG CM	NO OF TANKS	TANK CAPACITY	SEMI SYM SIZE	SEMI CUB SIZE	CYLINDRICAL SIZE	BI-LOBE SIZE
Cruise ship	3000	5100	4	1276		19x14x6.55	24x8	
Large Container	13000	22100	4	5525		CDTS 37x30x25	3 @ 50x17	
Great Lakes	860	1462	2	731	12.8x12.8x5.9	16.6x11.4x5.2	21.15x7.05	15.75x10x6.3
Coastal Ferry	51	87	1	87	6.5x6.5x2.75	8.35x5.9x2.45	10.5x3.5	7.63x4.88x3.05
Coastal Ferry		65	2	33	5.2x5.2x2.1	6.4x4.6x1.8	9x2.25	
Chemical Tanker	2300	3910	2	1955	15.8x15.8x7.4	20.95x14.3x6.65	26.4x8.8	19.75x12.64x7.9
Channel Ferry	1230	2091	2	1046	14.1x14.1x6.55	18.4x12.6x5.8	23.4x7.8	17.5x11.2x7.9
RO RO	5310	9027	2	4514	25.4x25.4x12.2	31x21x10	39x13	29.75x19x11.9
RO RO	1303	2215	2	1108	14.5x14.5x6.75	19x13x6	24x8	17.88x11.44x7.15
Tanker	1000	1700	2	850	13.5x13.5x6.25	17.5x12x5.5	22.5x7.5	16.5x10.56x6.6
OPV		12	1	12	3.5x3.5x1.25	4.45x3.3x1.15	5.25x1.75	3.88x2.48x1.55

5. STRUCTURAL ANALYSIS

Type 'C' LNG fuel tanks must meet both the IGF code as well as ASME pressure vessel requirements. A summary of the preliminary structural analyses per references (Wang, Shin & Norris, 2015) performed for a Type 'C' SEMI-CDTS fuel tank is described in the next section. Per IGF Regulations for Type 'C' independent tanks, the allowable stress limits are:

$\sigma_m \leq f$
$\sigma_L \leq 1.5f$
$\sigma_b \leq 1.5f$
$\sigma_L + \sigma_b \leq 1.5f$
$\sigma_m + \sigma_b \leq 1.5f$
$\sigma_m + \sigma_b + \sigma_g \leq 3.0f$
$\sigma_L + \sigma_b + \sigma_g \leq 3.0f$

where:
σ_m = equivalent primary general membrane stress;
σ_L = equivalent primary local membrane stress;
σ_b = equivalent primary bending stress;
σ_g = equivalent secondary stress; and
f = the lesser of R_m /A or R_e /B

The stresses σ_m, σ_L, σ_g and σ_b represent the stress responses due to membrane or bending load components. General membrane stresses are the nominal or far-field uniform-through-thickness in plane stress component while local membrane stresses are localized and occurring near constraints or discontinuities. Bending stresses are the nominal or far-field variable stresses across the thickness of the section minus the membrane component. The secondary stresses are the localized bending stresses including the thermally induced stresses.

R_e is the specified minimum yield stress at room temperature and R_m is specified minimum tensile strength at room temperature. The values A and B for various cryogenic grade metals have the following minimum (IMO, 2013):

	Nickel steels and carbon-manganese steels	Austenitic steels	Aluminum alloys
A	3	3.5	4
B	1.5	1.5	1.5

The material properties for 9% Nickel steel from reference (Smith & Craig, 2001) are as follows:

Property	T °C	9% Ni Steel
Yield stress, MPa	0	441-587 min specified 650(NNT), 690(QT)
	-196	680(NNT), 840 (QT)
Ultimate tensile stress, MPa	0	637-834 min. specified 840 (NNT)
	-196	1100(NNT), 940(QT)
Elongation to break, %	0	17-20 min. specified 30
	-196	20

Using the above formulas and material properties, the allowable stresses for 9% Nickel steel are as follows:

	at 0 °C	
IGF Stress Allowables	Nominal	Local
Allowable Membrane Stress (MPa) (i.e., structural segments that are bodies-of-revolution under normal pressure loading or plate structures under in-plane loading)	212	318
Allowable Total Stress (primary+secondary, incl. thermal) (MPa) (i.e., structural segments that are subjected to membrane and bending loads)	318	637

Altair's HyperWorks suite of structural analyses software was used to optimize a 1.0K Type 'C' Semi-CDTS tank to meet the required load cases. Since the proposed design is in the early development phase, in-lieu of a full sea-keeping assessment to model the dynamic motion of the vessel during its operating life under 10^8 North Atlantic wave encounters, the dynamic loads caused by the ship motion is first estimated using IGC rules (ABS, 2013) per ABS guidelines.

The calculated accelerations are based on the tank center of gravity. a_{le} , a_{te} and a_{ve} are the maximum dimensionless accelerations (i.e., relative to the acceleration of gravity) in the longitudinal, transverse and vertical directions, respectively and they are considered as acting separately for calculation purposes. a_{ve} does not include the component due to the static weight, a_{te} includes the component due to the static weight in the transverse direction due to rolling and a_{le} includes the component due to the static weight in the longitudinal direction due to pitching. The following formulas (ABS, 2013) are given as guidance for the components of acceleration due to ship's motions in the North Atlantic:

Vertical acceleration:

$$a_{ve} = \pm a_o \sqrt{1 + \left(5.3 - \frac{45}{L}\right)^2 \left(\frac{x'_T}{L} - 0.45\right)^{1.5}}$$

Transverse acceleration:

$$a_{te} = \pm a_o \sqrt{0.6 + 2.5\left(\frac{x'_T}{L} - 0.45\right)^2 + K\left(1 + 0.6K\frac{y_T}{B}\right)^2}$$

Longitudinal acceleration:

$$a_{le} = \pm a_o \sqrt{0.06 + A^2 - 0.25A}$$

With

$$A = \left(0.7 - \frac{L}{1200} + 5\frac{y_T}{L}\right)\left(\frac{0.6}{C_b}\right)$$

Where:

a_o = $k_p\left[0.2\left(\frac{V}{\sqrt{L}}\right) + \frac{(34 - 600/L)}{L}\right]$

k_p = load factor for adjusting the probability of exceedance

= 1.0 for yielding and buckling strength assessment

0.5 for fatigue strength assessment

x'_T = $x_T - (L_{BP} - L)$

K = 1.0 in general. For particular loading conditions and hull forms, determination of k according to the formula below may be necessary:

= $13GM/B$, where $K \geq 1.0$

GM = metacentric height

V = 10 knots for yielding and buckling strength assessment

= 75% of design speed for fatigue assessment

Note: The accelerations calculated from the recommended formula in IGC Code at 10 knots are considered representative of the ship's actual accelerations in extreme sea conditions.

Yield Load Cases

The ABS dynamic load criteria used is based on the long-term extreme value for North Atlantic waves corresponding to a probability of exceedance of 10^{-8}. For a vessel with the following principal dimensions: L_{BP}= 250 m, Beam = 49 m, Depth = 25 m, L_{OA} = 260 m, the longitudinal acceleration, transverse acceleration and vertical acceleration are: a_x= 0.12g, a_y =0.27g, a_z= 0.70g, p=0.5MPa. Corresponding load components for load cases 1 to 4 are summarized in Table II:

Table II. Yielding load cases summary

Load Case	Longitudinal acceleration	Transverse acceleration	Vertical acceleration	Gravity load	Design pressure
1	a_x	$-a_y$	$-a_z$	$-g$	p
2	a_x	$-a_y$	a_z	$-g$	p
3	$-a_x$	$-a_y$	$-a_z$	$-g$	p
4	$-a_x$	$-a_y$	a_z	$-g$	p

Envelope stress contour plots (in MPa) for the 4 load cases are shown in Figure 5 and Figure 6.

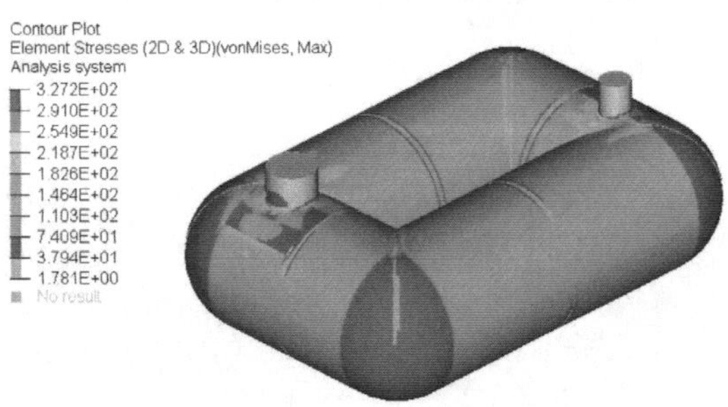

Figure 5. Yielding Load Case Envelop Stress Result

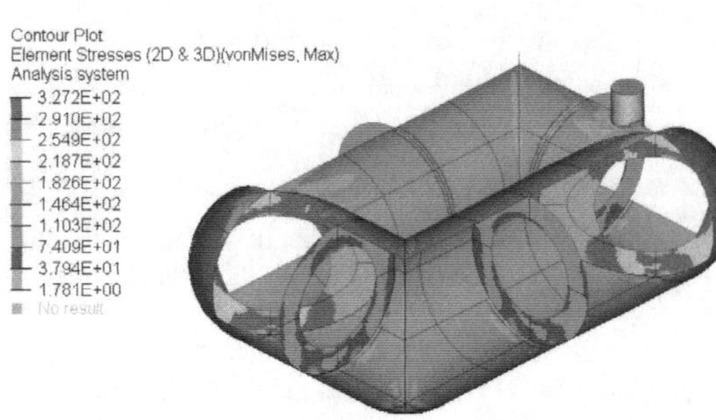

Figure 6. Yielding Load Case Envelop Stress Result

Fatigue Load Cases

The ABS dynamic load criteria used is based on the long-term extreme value for North Atlantic waves corresponding to a probability of exceedance of 10-4. Load pairs for fatigue load cases were assessed per reference (ABS, 2013). Results of the fatigue analyses which were done per the methodology outlined in reference (Ramoo, *et al*, 2016) for a CDTS tank, are not included in this paper for brevity.

Accidental Loads

Per IGC code, a forward (Load Case 5) and aft collision (Load Case 6) and a flooding (Load Case 7) were evaluated to ensure that the self-centering support arrangement can withstand these peak loads. Stress contours (in MPa) corresponding to these load cases are shown in Figures 7, 8 and 9.

Table III. Accidental load cases

Load Case	Longitudinal acceleration	Lateral acceleration	Vertical acceleration	Gravity load	Design Pressure
5	0.5g	0	0	$-g$	p
6	-0.5g	0	0	$-g$	p
7	0	0	0	$-g$	p

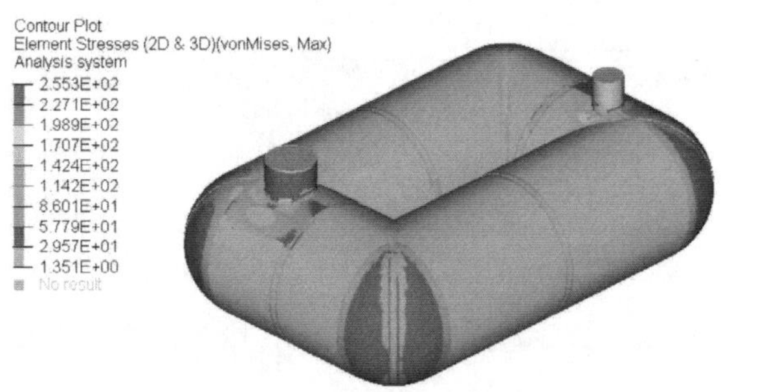

Figure 7. Front Collision Result

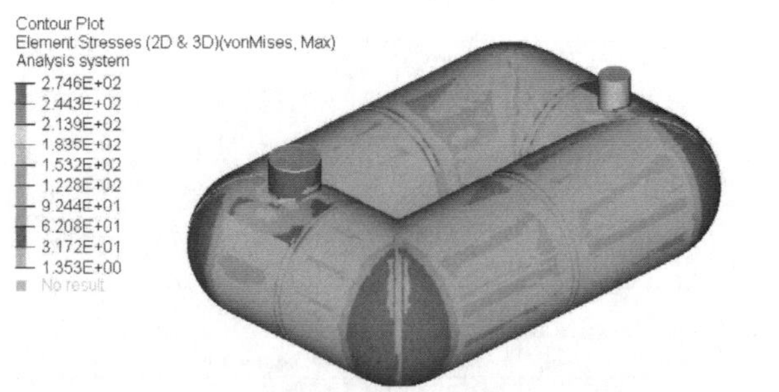

Figure 8. Rear Collision Result

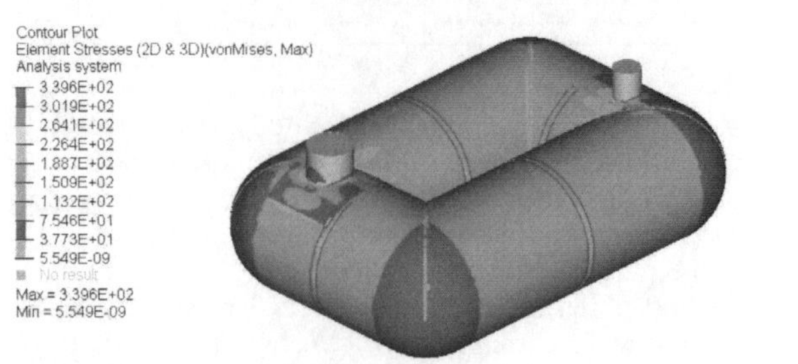

Figure 9. Flooding Condition Result

Buckling Loads

An eigenvalue analysis was performed to assess the buckling strength of the shell as well as the ring webs. Though more applicable for a below deck installation, for a preliminary evaluation, the tank buckling propensity was evaluated (Figure 10) by assuming the tank was empty and submerged at a depth equivalent to the tank diameter. The buckling margins can be increased by stiffening up the region where the lowest eigenvalue is seen.

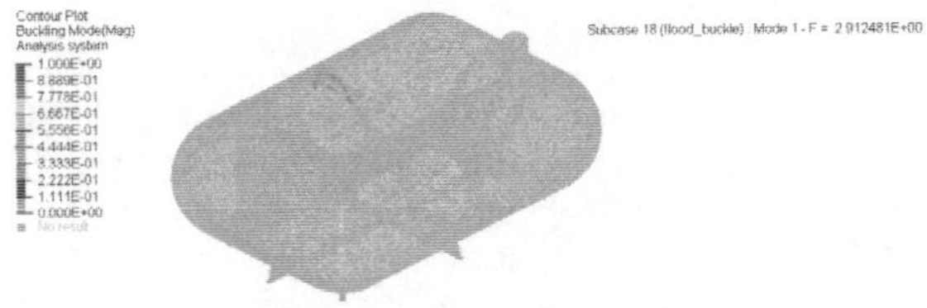

Figure 10. Buckling Analysis

Sloshing Loads

A sloshing evaluation was done using Altair's RADIOSS software. The simulation methodology was based on the ALE (Arbitrary Lagrangian Eulerian) approach to expediate the initial preliminary assessment of sloshing loads. The sloshing impact loads were assessed for multiple fill levels and 3 wave headings. The sloshing simulation was done per the approach outlined in references (Ramoo *et al*, 2016, Wang, Shin & Norris, 2015). The sloshing pressures were estimated assuming a conservative worst-case excitation period of 6 seconds. The highest sloshing pressures were from bow quartering waves at about 60% fill level. The input excitation and corresponding output pressures at the highest location is shown in Figure 11. As per the design intent, the bulkhead ring webs and corners ring webs, attenuate the sloshing waves by ensuring that the annular depths of these structures are sufficient to space the containment system resonant periods away from the vessel's roll and pitch periods. The ALE model used for the simulation is shown in Figure 12.

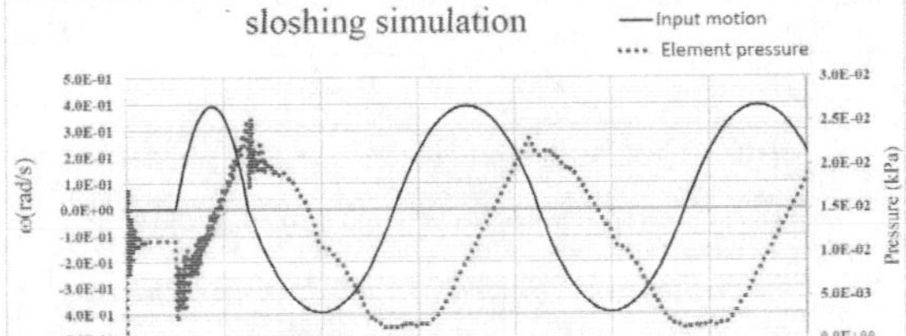

Figure 11. Roll Motion with a Period of 6s

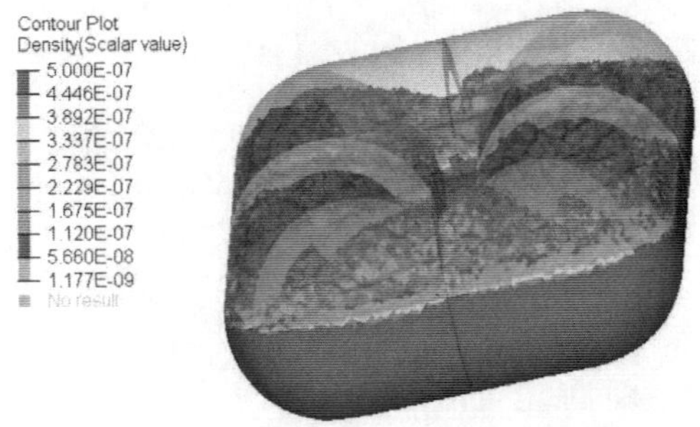

Figure 12. Sloshing Model

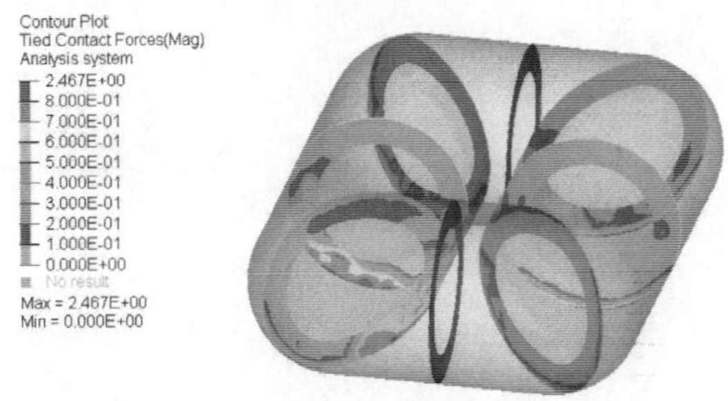

Figure 13. Sloshing force contour (t=5.6s)

Thermal Contractions due to Cryogenic Cargo

A thermal contraction evaluation was performed to ensure that the self-centering support arrangement allows for the tank to freely contract yet providing a means to prevent excursions due to the vessel motions including extreme accelerations as described by the accidental conditions. The contour plot below shows the tank contractions (in mm) due to loading of the cryogenic cargo.

The corresponding near zero stress state (in MPa) on the supporting structure shows how the four support-block arrangement allows for free transverse and longitudinal contractions. The use of four sliding guide supports creates a self-centering arrangement about a virtual central constraint point.

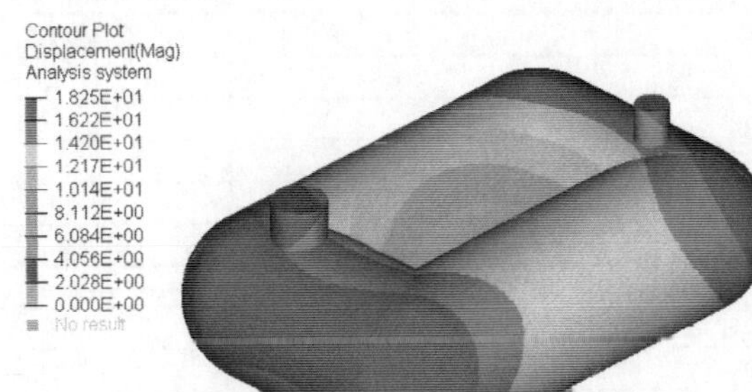

Figure 14. Thermal Contraction Displacement

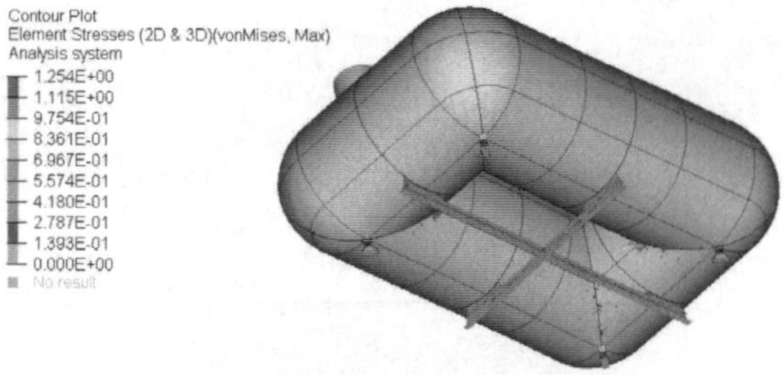

Figure 15. Thermal Contraction Stress

Figure 16. 7.5 K CM LNG Bunkering Vessel

6. APPLICATIONS

6.1 COASTAL AND BUNKER VESSELS

A wide range of small LNG Carriers have been built and many more planned for coastal service and/or bunkering vessels for LNG fuelled ships. They range from 2,500 to 50,000 CM. The most important characteristic for these ships is their draft. Being coastal and harbour vessels, draft can be restricted from 3.75 to 7 m. Thus far, they have been built using Cylindrical and Bi-lobe tanks. Designs have also been proposed for membrane and prismatic tanks. Table IV presents a design for a 7,500 CM vessel including all tank types including the SEMI-CDTS. From the Table, the advantages of the SEMI are again clear. Note that the cost estimates are given for comparison value only as percentage differences between designs. The cost will differ depending on shipyard productivity, labor rate, overhead, and profit, and even monetary exchange rates.

The price will depend on market conditions both local and international and often does not include any profit. Figure 16 is a Concept General Arrangement for the 7,500 CM with three SEMI-CDTS tanks.

Other possibilities for the SEMI-CDTS storage and bunkering vessels are conversions of Offshore Supply Vessels (Figure 17) and barges (Figure 18).

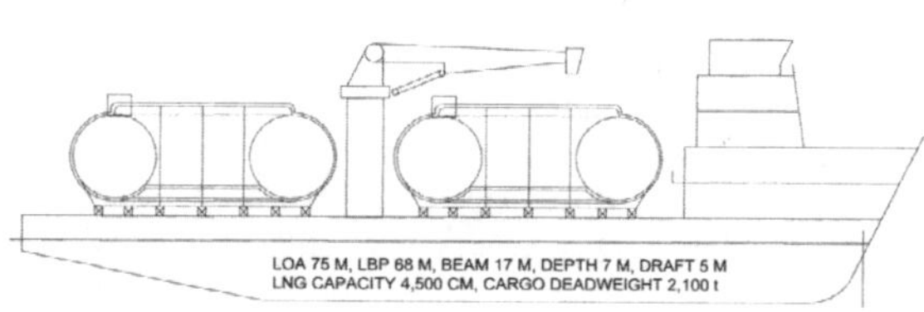

Figure 17. OSV Conversion to LNG Bunker Vessel

Table IV. Comparison of 7.5 k cm lng bunker vessels

CHARACTERISTIC		SEMI STEEL TYPE B	SEMI STEEL TYPE C 4.5 BAR	CYLINDRICAL TYPE C 4.5 BAR	BI-LOBE TYPE C 4.5 BAR	MEMBRANE 2 BAR
LOA	m	112.5	113.5	118.4	120	115
LBP	m	108	108	111.7	117	110
Beam	m	21	21	18.6	19.5	20
DEPTH MAIN DECK	m	9	9	9.2	11.6	9.4
DEPTH COVER	m	12.5	12.5	14.1		12.9
DRAFT	m	4.73	4.95	5.55	5	5
SPEED	Knots	13	13	13.5	13	12
DEADWEIGHT	t	3590	3620	3583	3647	3602
SHP	kW	2433	2494	3456	3152	2129
Building Cost First Ship w Engineering		X-14%	X	X+9%%	x+4%	X+12%

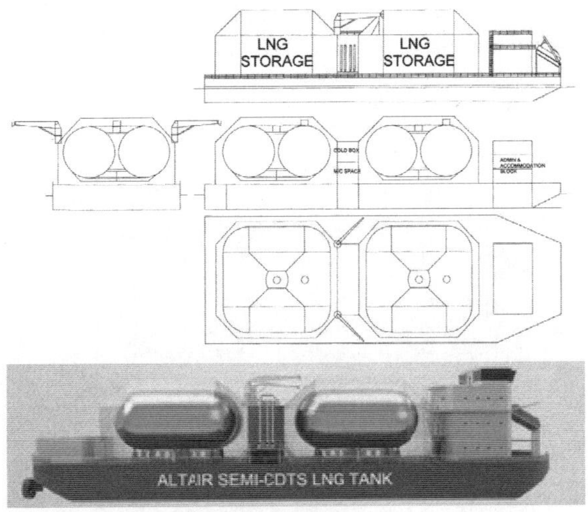

Figure 18. 5 K CM Storage Barge

6.2 FUEL TANKS

LNG versus MDO as a Fuel
One of the disadvantages of LNG as a fuel is that LNG has about 60 percent of the Btu of marine diesel for equal volumes, thus requiring about 1.7 times the containment space of MDO. Another is that it needs to be stored at low temperature or high pressure, and due to heat ingress, evaporates gas (boil off) over time. As most ships operate at different machinery loads during their normal operation this boil off gas needs to be either contained in the tank, re-liquefied or burnt in a GCU as a last option. The first option requires the tanks to be designed as pressure vessels rated to the required holding period and pressure build up. The second option allows for a Type B (rated up to 0.7 bar) but then requires a re-liquefaction plant to be installed on the ship taking up more space, depending on the use and rate of use of the boil off gas. For most current applications, the operators have chosen Type C tanks.

Conversion Using Existing Space or Deck
For conversions, there are usually two methods proposed. Both need to take the ship out of operation to install LNG tanks and fuel handling systems in the existing hull. Such conversions will require the use of existing space that is currently in use for other required service. One is to take the ship out of operation to install LNG tanks and fuel handling systems in the existing hull or on an open space on deck. The second is to install a newly prefabricated hull section (mid-body or stern) that contains all the necessary LNG systems. The new hull section can be completed before the ship is taken out of service. A benefit is that the new hull section can provide additional space that can be used for other earning services.

All conversions can be time-consuming and will cause short-term loss of revenue by removing the ship from service. Obviously, it is less time and work effort if the tank(s) can be installed in an open space on a deck or in an existing hold, compared to the insertion of a new hull section, which is the most expensive in both time out of service and cost

Small Craft
Many Tugs, Offshore Supply Vessels, Research Vessels, and Offshore Protection Vessels are being built or converted to LNG fuel. For many of these small crafts, tank length, height, floor print, and for stability, vertical center of gravity, are all restricted. Each application is a custom solution. The SEMI-CDTS tank is an ideal application for these vessels.

Figure 19 shows a 12 CM tank in the design compartment allocated for the LNG tank. Neither the Cylindrical or Bi-lobe could fit into the available space (insufficient deck height).

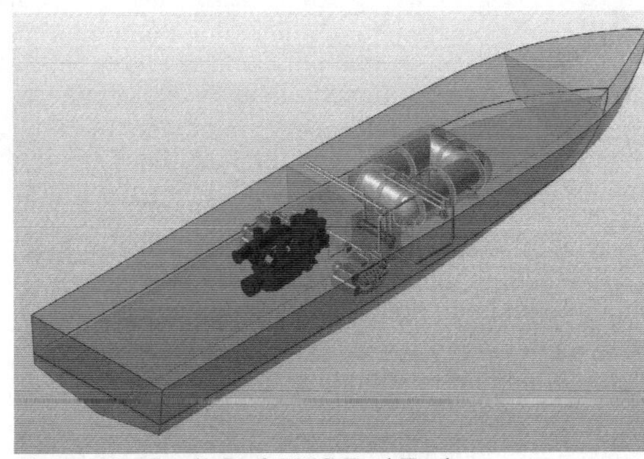
Figure 19. Small Craft LNG Fuel Tank

Ferry
Many ferry companies have been quick to announce LNG conversions. In North America, these include operators as diverse as BC Ferries, Washington State Ferries, and the Staten Island Ferry, all of which operate large day boats, while in Europe, several newly-built LNG-powered ferries are already in operation, chiefly in Scandinavia.

A study to convert an existing car ferry to duel fuel engines proposed two tanks 12 m x 4.2 m diameter to provide 180 CM. Figure 20 shows how two SEMI-CDTS Cuboid tanks, each 12 m x 5 m x 2 m would fit in same location. The SEMI-CDTS Cuboid tank had to fit in between two stacks, hence its breadth was restricted resulting in a lengthened SEMI-CDTS Cuboid Tank, thus showing the flexibility of the approach. The reduction in weight of the SEMI-CDTS Cuboid tank and its vertical center of gravity by 1.1 m has a significant beneficial impact on the stability of the Ferry.

Cargo Ship
Cargo ships come in many varieties and because of this there is no single best solution for the location of the LNG fuel tanks. On all aft deckhouse ships it would be

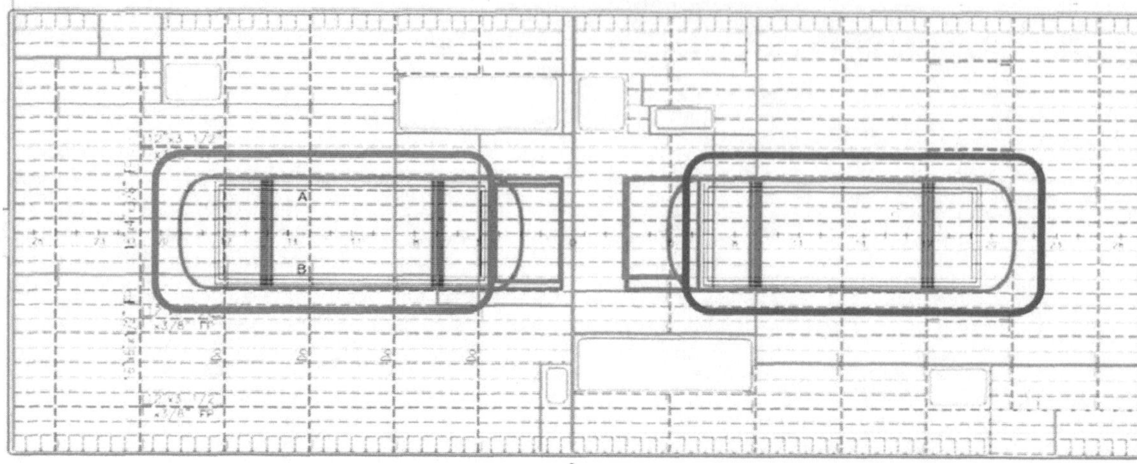

Figure 20. Comparison of Cylindrical Tanks and SEMI Cuboid Tanks for a Car Ferry

convenient if the tank(s) could fit aft of the deckhouse on deck. The space usage efficiency of the SEMI-CDTS tank provides more cost effective design solutions.

Tanker

Most Tankers will install the LNG tanks on deck as in the recent reported conversions of several Tankers. Figure 21 is a comparison of Cylindrical tanks and SEMI-CDTS. The vertical center of gravity of the tanks will be reduced by 1.25 m when using the Cuboid SEMI-CDTS tank at the same height off deck for mooring line clearance and a further 3 m CG height reduction by mounting the SEMI-CDTS tank on deck in the space forward of the mooring line access. The transverse fuel tank positioning must meet the probabilistic side distance requirement of $f_{CN} < 0.04$, as an alternative to the requirement of the distance from the side being the lesser of 11.5 m or B/5 measured inboard from the ship side at right angles to the centreline at the summer draft load line (IMO, 2013).

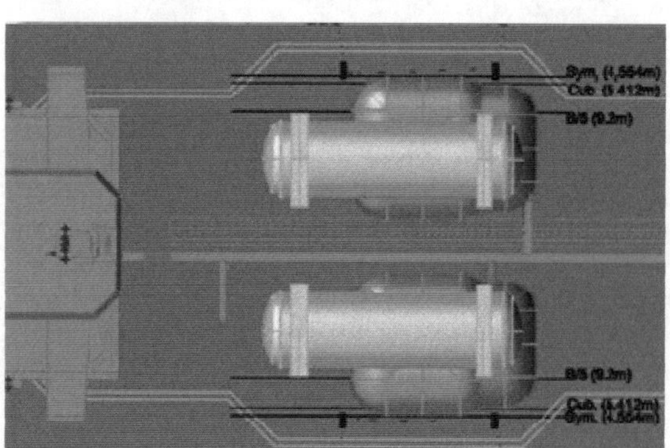

Figure 21. Safe Tank Location per IGF Code

Container Ship

Container ships are MHO propelled with smaller tanks storing MDO for manoeuvring and port operations. LNG would replace the use of MHO for voyage propulsion. A number of locations have been proposed for the installation of LNG fuel tanks on container ships for new builds as well as conversions. TOTE has new builds and conversions where the tanks are located aft of the deckhouse on the weather deck. For the largest container ships, locating some of the required tanks in the hull below the deckhouse offers the least impact on container capacity. Locating tanks, in a new hull section, to be inserted into the existing ship can add space for deck containers. For large container ship conversion, one CDTS tank or two SEMI-CDTS tanks can be used. Figure 22 shows how CDTS tanks would compare to cylindrical tanks. While combination RO RO/Container ships in service have the tanks all the way aft and on deck, membrane tank designers have proposed an under deckhouse plus one or more container bay location, option 1. This location would eliminate 616 TEU. This however, is an unfair comparison to other arrangements/tanks as such arrangement assumes that existing systems in the space below the deckhouse would not have to be located somewhere else on ship. A fairer comparison would be to have the Membrane and Prismatic tanks located in container bays, as shown in option 2 where 924 TEU would be eliminated. The cylindrical tank, option 3, would have 3 tanks aft on deck. This would eliminate 1252 TEU. The CDTS has two options; option 4 has the tank located aft of the machinery space, partially in the aftermost holds and extending above deck, which eliminates 608 TEU, if a deck is constructed over the tank to carry containers; and option 5, in the holds forward of machinery space location, which eliminates 924 TEU the same as option 2 for membrane and prismatic tanks. Option 6 uses two SEMI-CDTS tanks in separate compartments, one forward and one aft of machinery space and 840 TEUs would be eliminated.

Self-standing tank-in the hull installations would be the least cost and time out of service. In the hull membrane tanks would take the most time. Option 1 is also furthest from the machinery space and would require running double walled and insulated piping through container holds. All the other options are clustered around the machinery space thus shorter piping distances.

Bulk Carrier

An interesting study was carried out by a consortium, The Green Corridor Bulker, proposed two cylindrical tanks, each 3000 CM, in separate compartments and below the accommodation. Figure 23 shows the proposed arrangement. The use of a single SEMI-CDTS tank in the center using an offset machinery room uptake results in better space utilization. As per tankers, fuel tank positioning must meet B/5, or the probabilistic side distance requirement of $f_{CN} < 0.04$ (IMO, 2013).

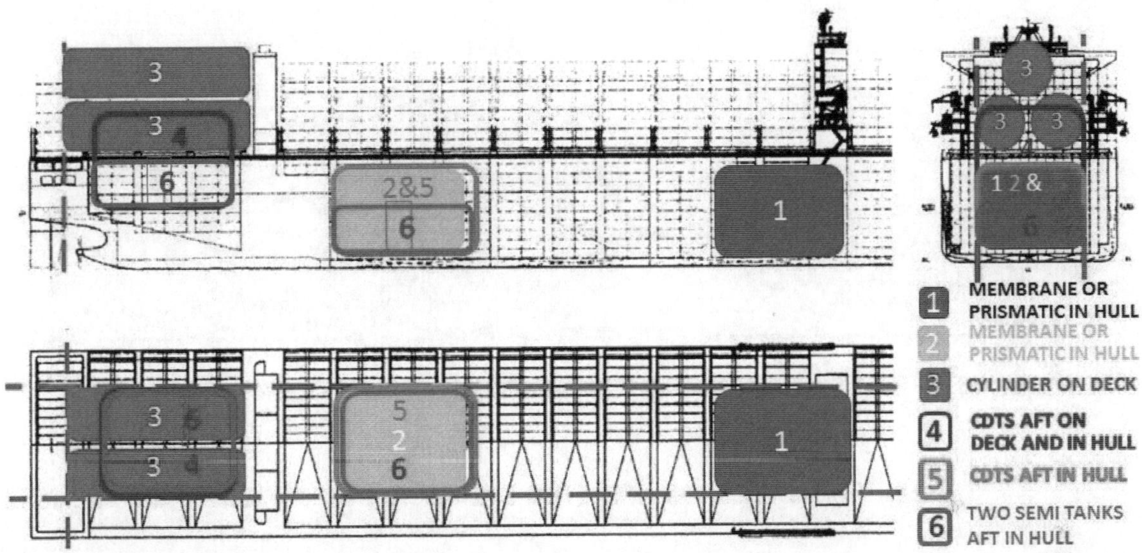

Figure 22. Tank Locations for Container Ship

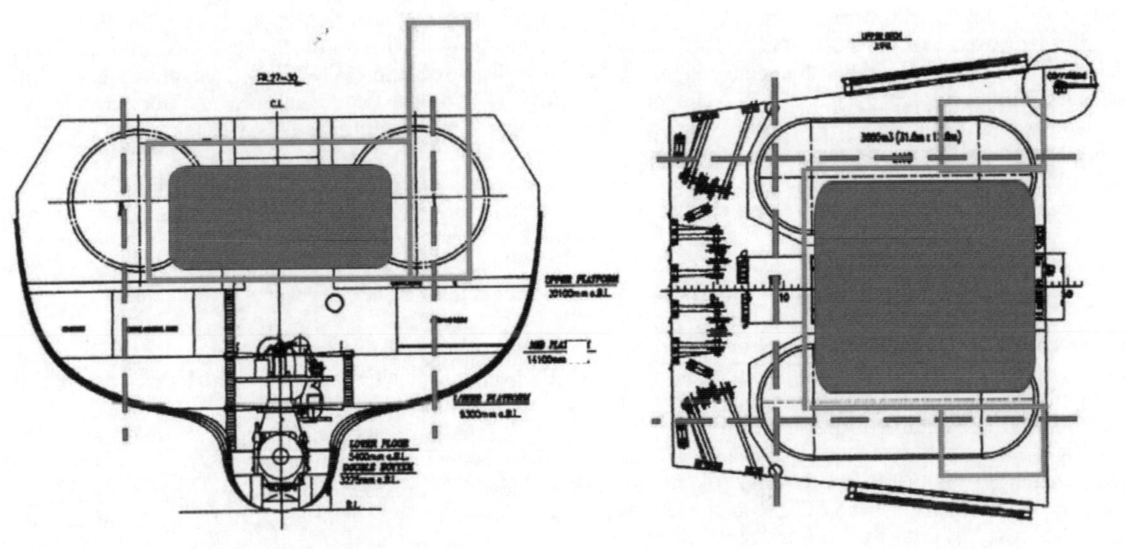

Figure 23. Green Corridor Bulker SEMI Application

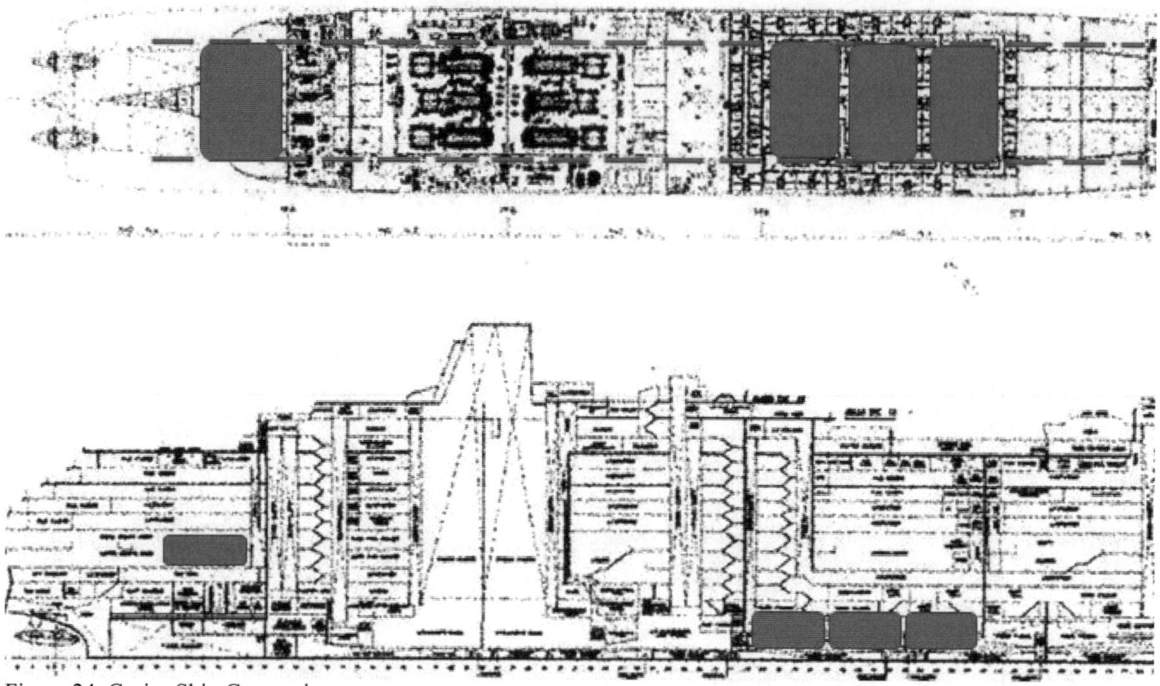

Figure 24. Cruise Ship Conversion

Cruise Ship
Because of their time near land and important tourist land sites, cruise ships are a natural LNG fuel candidate to reduce harmful emissions from marine fuel oil. However, they are possibly the most challenging regarding locations for conversions as most spaces are restricted by deck height and close spaced transverse bulkheads. It is obviously easier to design LNG tanks into new cruise ships as they are being built, but the conversion of existing cruise ships is under consideration by many cruise ship operators.

Conversions will either require the use of existing space that is currently in use for other required service, that is reduce the number of cabins, service compartments, and possibly deck space, or a new hull section containing the LNG fuel tanks can be inserted into the existing ship. The latter approach offers the ability to add to the number of passengers that can be carried by locating cabins in the new section in turn offering increased future revenue to help pay for the conversion. Figure 24 shows a cruise ship conversion approach that has 12 MDO fuel tanks replaced by 4 identical SEMI-CDTS LNG fuel tanks.

7. CONCLUSIONS

The SEMI-CDTS tank offers a competitive alternative to the existing Cylinder and Bi-lobe tanks for small scale applications. They provide flexibility in shape, and significant reduction in height and length, thus the ability to fit into existing compartments for conversions compared to cylindrical and bi-lobe tanks.

The unique design allows for a modular build since each quarter section is identical. The use of smaller diameter and shorter cylindrical segments, and a more evenly distributed support arrangement reduces stress concentrations at the critical support locations.

The SEMI-CDTS offers:
- weight reduction; up to 10%/13% compared to cylindrical/bi-lobe tanks
- reduction in tank height, up to 28%/17% compared to cylindrical/bi-lobe tanks
- lower vertical center of gravity
- reduction in draft up to 20% for LNG Bunker and Coastal vessels

- reduction in building cost up to 10% to 20% compared to cylindrical, bi-lobe and membrane tanks
- flexibility of location as fuel tanks
- location of fuel tanks in container ships that reduces eliminated containers by up to 30%
- smaller cylindrical segment diameters and a modular build

8. ACKNOWLEDGEMENTS

The authors would like to acknowledge with thanks the support of ALTAIR Engineering and their vision of a future for the CDTS. The application of their advanced CAE tools to a practical idea has enabled the CDTS to be further developed and proven its feasibility.

9. REFERENCES

1. LAMB, T, and RAMOO, R, '*The Application of a New Tank Containment System to ULTRA-Large LNG Carriers*', OTC 2009a
2. LAMB, T, and RAMOO, R., '*A New Concept for CNG Carriers and Floating CNG/Oil Processing and Storage Offshore Platforms*,' CNG Forum, London 2009b
3. LAMB, T, and RAMOO, R, '*A New Tank Containment System for Floating LNG and CNG Processing and Storage Offshore Platforms*,' OTC 2010
4. LAMB, T, and RAMOO, R, '*An Economic Study of LNG Carriers with Different Containment Systems on Specific Trade Routes*,' RINA Design & Construction of LNG Ships, London, 26-27 October 2016
5. RAMOO, R, ENG, B.W, WU, S. H and LAMB, T, '*The Optimum configuration of a Self-Standing Type B LNG Tank using Advanced Structural Analysis and Simulation Tools*,' RINA Design & Construction of LNG Ships, London, 26-27 October 2016
6. WANG, B, SHIN Y. S, NORRIS, E, '*Strength Assessment of Type 'C' LNG Fuel Tanks*,' Proceedings of the ASME 2015 34th International Conference on Ocean, Offshore and Arctic Engineering, May21-June 15, St. John's, Newfoundland, Canada
7. IMO IGF Code: *International Code of Safety for Ships using Gases or other Low Flashpoint Fuels*, 2013
8. IMO IGC Code: *International Code for the Construction and Equipment of Ships Carrying Liquefied Gases in Bulk*, 1993.
9. ASME *Boiler and Pressure Vessel Code*, Section VIII, Division 2, 2013.
10. American Bureau of Shipping (ABS), *Rules for Building and Classing Steel Vessels, Part 5C*, 2013.
11. American Bureau of Shipping (ABS), *Guide for Building and Classing Liquefied Gas Carriers with Independent Tanks*, 2010.
12. SMITH, L., CRAIG, B., '*Properties of Metallic Materials for LNG Service*', 9th MECC, 12th-14th, Bahrain, 2001.

AUTHORS BIOGRAPHY

Mr. Lamb has over 60 years' experience in the design and construction of many ships of diverse types and spent his last 11 active years teaching and performing research at University of Michigan, retiring in 2006. He served a five-year apprenticeship at H.M. Dockyard Rosyth and graduated from Durham University in 1958 with his B.Sc. in Naval Architecture. He is a Fellow of RINA, and SNAME, and was the Editor of, and 5-chapter contributor, to the 2003 edition of the SNAME SHIP DESIGN AND CONSTRUCTION book as well as being the SNAME Ship Design Committee Chairman for many years. He is a Professional Engineer in the U.S., Chartered Engineer in UK and Eur Ing.

Mr. Ramoo, Vice President of Engineering at Altair ProductDesign has over 30 years' experience in the use of computational methods for design ideation and Advanced Structural Analysis. Mr. Ramoo graduated with a Diploma in Marine Engineering from the Singapore Polytechnic in 1977 and holds a MSc. in Mechanical Engineering from The Ohio State University. He has broad experience in using advanced CAE methods in structural design of land vehicles and marine structures.

Mr. Eng, Engineering Manager at Altair Taiwan has over 10 years' experience in the use of computational methods in structural analysis. Mr. Eng graduated with a MSc. in Engineering Science and Ocean Engineering from The National Taiwan University in 2005. His experience is in heavy industry and ship structure design using advanced CAE methods.

Mr. Egambaram, Senior Project Engineer at Altair Malaysia has over 8 years of experience working with process, mechanical & piping engineering and design in the shipbuilding and oil & gas industries. He holds a B. Eng (Hons) in Manufacturing Design from The Technical University of Malaysia.

Mr Nadzri, Project Engineer at Altair Malaysia works with Computer Aided Design, in the field of Naval Architecture. He holds a Diploma in Engineering Technology specializing in Ship Design from the University of Kuala Lumpur.

LNG POWERED DRY BULK CARRIER

A Trakakis, Arista Shipping, Greece, **M Ioannou**, **M Penfold** and **E Kariambas**, American Bureau of Shipping, Greece and UK and **D Antonopoulos**, Wärtsilä, Greece

SUMMARY

LNG is one of the potential alternative fuel solutions to meet the specified IMO fuel sulphur content limits since it contains negligible sulphur. Furthermore, gas and DF engines utilizing the Otto combustion process can produce NOx emissions below the IMO Tier III limits. This paper discusses the challenges involved and how these were met, when considering LNG as fuel in an ocean-going dry bulk carrier, with focus on the main engine technologies and machinery, and their effect on the vessel design. The study was conducted by Arista Shipping, the American Bureau of Shipping (ABS) and Wärtsilä, and was based on a representative bulk carrier design, an 82,000 dwt Kamsarmax vessel, which was selected as the project platform due to its proven design and high cargo capacity.

NOMENCLATURE

2S	2-stroke (engine)
4S	4-stroke (engine)
BoG	Boil off Gas
CAPEX	Capital Expenditure
CFD	Computational Fluid Dynamics
CPP	Controllable Pitch Propeller
dwt/DWT	Deadweight tonnage
DF	Dual Fuel
ECA	Emission Control Area
EEDI	Energy Efficiency Deign Index
EGCS	Exhaust Gas Cleaning System
EU	European Union
FGSS	Fuel Gas Supply System
GHG	Green House Gases
HFO	Heavy Fuel Oil
HP	High Pressure
HSFO	High Sulphur Fuel Oil
IGC	International Code for the Construction and Equipment of Ships Carrying Liquefied Gases in Bulk
IGF	International Code of Safety for Ships Using Gases or Other Low-flashpoint Fuels
IMO	International Maritime Organisation
LNG	Liquefied Natural gas
LP	Low Pressure
LRM	Light Running Margin
LSFO	Low Sulphur Fuel Oil
mbarg	millibar gauge (unit of pressure) mmBTUThousand thousand (i.e. million) British Thermal Units
MCR	Maximum Continuous Rating
MEP	Mean Effective Pressure
MEPC	Marine Environment Protection Committee
MGO	Marine Gas Oil
MMBTU	Million British Thermal Units
MN	Methane Number
MT	Metric Ton
NOx	Nitrogen Oxides
OPEX	Operating Expense
PTO	Power Take Off
SOLAS	Safety of Life at Sea
SOx	Sulphur Oxides
THC	Total Hydrocarbons
USD	United States Dollars.

1. INTRODUCTION

In recent years, marine engine exhaust emissions legislation has become increasingly stringent. The upcoming global fuel sulfur content limit of 0.5%, effective from 1st of January 2020, together with the existing ECA limits for fuel sulfur content of 0.1% and the Tier III NOx limits, impose significant challenges to the shipping industry. The options available to achieve compliance critically involve the selected engine technology and type(s) of fuel, two choices which then determine whether exhaust emission abatement technologies are necessary.

LNG contains negligible sulphur and it therefore meets all IMO regulations with regard to SOx emissions, as controlled by the fuel sulphur content limits. Furthermore, when combusted in gas or DF engines operating according to the Otto combustion cycle NOx emissions below the IMO Tier III limits are possible. On a calorific value basis, LNG prices are lower than traditional marine fuel prices. However, there are still challenges for the application of LNG as a fuel, notably fuel containment, fuel preparation systems, additional CAPEX and the LNG bunkering infrastructure. Given its strong potential, and despite the anticipated challenges, the use of LNG as a fuel is now extending beyond the application to LNG carriers, to other ship types such as bulk carriers and tankers.

This paper discusses the challenges involved in considering LNG as fuel in an ocean-going dry bulk carrier employed on varied international shipping routes and focuses on the main engine technologies and machinery space arrangements. The study is based on a bulk carrier design, an 82,000 dwt Kamsarmax vessel.

The study concentrated on the selection criteria and available technologies for the main engine and the respective machinery space arrangements, whilst taking into account the applicable Classification requirements, relevant statutory safety regulations (IGF code) and the shipowners operational requirements to deliver a highly efficient competitive solution, with increased reliability and redundancy characteristics.

The findings show that despite the higher CAPEX incurred for the application of LNG as a fuel, the payback period could potentially be a few years and provide a step forward in delivering a competitive, environmentally friendly merchant vessel with favourable total cost of ownership.

The study took into account all major factors that affect the selection of the propulsion system and the vessel design. Such factors included consideration of the applicable emission legislation, the various types of LNG containment systems and the location on board, together with the impact (stresses/stability) on the vessel design, the machinery requirements and the type of main engine to be adopted. Further considerations were given to CAPEX and OPEX, as well as to fuel prices, the fuel supply infrastructure, and to the regulatory framework. The LNG fueled bulk carrier design discussed in this paper received an Approval in Principle from ABS on 22nd of August 2016.

2. IMO SAFETY REGULATION

The IMO adopted the *International Code of Safety for Ships Using Gases or other Low-flashpoint Fuels (IGF Code)* by Resolution MSC.391(95) in June 2015. At the same time Resolution MSC.392(95) was adopted implementing amendments to SOLAS to make the IGF Code mandatory, introducing a new 'Part G' for ships using low flashpoint fuels and applying an entry into force date of 1 January 2017. The basic philosophy of the Code is to provide mandatory provisions for the arrangement, installation, control and monitoring of machinery, equipment and systems using low-flashpoint fuel to minimize the risk to the ship, its crew and the environment.

The IGF Code provides detailed prescriptive safety requirements for ships using natural gas as fuel under Part A-1 of the Code. This prescriptive approach is based largely on the extensive experience gained with the carriage and burning of LNG as fuel in gas carriers covered by the IMO IGC Code over many decades. It also applies reduced risk assessment obligations for the use of natural gas as fuel.

The majority of the existing gas fuelled fleet experience (non gas carriers) has been gained with ships built to the preceding IMO Resolution MSC.285(86) *Interim Guidelines on Safety for Natural Gas-fuelled Engine Installations in Ships*, which was adopted 1 June 2009. However, the LNG powered dry bulk carrier detailed in this paper has been assessed on the basis of compliance with the IMO IGF Code.

3. IMO EMISSIONS REGULATION

There are two main drivers pushing the consideration of LNG as fuel for marine propulsion systems. The first relates to air pollution emissions legislation and the second to the availability and price uncertainty of conventional liquid fuels in the future. IMO Annex VI limits sulphur emissions and the sulphate fraction of particulate matter through the fuel sulphur content limits of Regulation 14 and NOx emissions through Regulation 13. Both regulations have global limits and also local limits regulated through Emission Control Areas (Figure 1). With the adoption of GHG emission controls through the energy efficiency requirements adopted by IMO MEPC 62 in July 2011 there is now also the GHG incentive for the selection of low carbon fuels such as LNG. Dual fuel and single gas fuel engines have the potential to reduce CO_2 emissions.

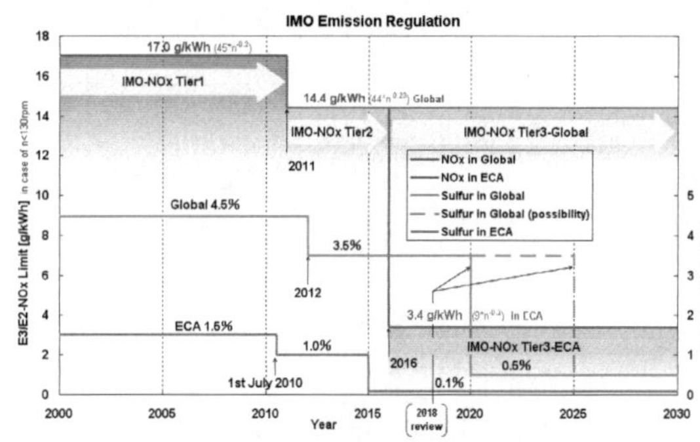

Figure 1. IMO Emissions Regulation (NOx and SOx)

Despite the recent low fuel oil prices, there remain long term question marks over future fuel prices, since a sudden shift in the demand of fuel type globally, from HSFO to 0.5% Sulphur fuel (Shell, 2017), estimated at a total of about 200 mil tonnes per annum, will have significant but not fully known consequences on the fuel bunker markets

On the other hand, the global abundance of gas, as epitomized by the unconventional North American shale gas supplies, provide the potential to deliver low cost LNG as a marine bunker fuel. This supply driver, along with the spectre of market based measures penalizing high GHG fuels, suggests that LNG can play a growing part in the marine fuel supply chain of the future.

3.1 ROUTES TO NOx COMPLIANCE

Shipowners, engine designers, builders or operators have three basic routes available to IMO Tier III NOx compliance:

- Conventional but advanced diesel engines running on either liquid or gas fuel using internal measures such as Exhaust Gas Recirculation (EGR);

- Conventional diesel engines running on liquid fuel using measures such as Selective Catalytic Reduction (SCR) after-treatment systems, which could be of the low pressure (LP SCR) or high pressure (HP SCR) type; or,
- Low pressure DF or gas engines operating according to the Otto combustion cycle that produce NOx emissions below the Tier III limits.

Each compliance option has its own advantages, disadvantages and CAPEX/OPEX impacts, which means the most suitable option is very much a ship and route/operating profile specific decision.

3.2 ROUTES TO SOx COMPLIANCE

The IMO fuel sulphur limit of 0.5% is applicable globally from 1 January 2020 and the 0.1% fuel limit in ECAs has been effective since 1 January 2015. Compliance with these SOx regulations is possible through two basic routes:

- The use of sulphur compliant fuels, which would include conventional liquid fuels such as MGO and other low sulphur fuels such as LNG; or,
- The use of Exhaust Gas Cleaning Systems (EGCS) on-board the vessel that reduce the exhaust sulphur emissions to an equivalent level as the fuel sulphur limits, when running on higher sulphur content fuels, typically using a wet exhaust scrubbing process.

The CAPEX and OPEX costs associated with EGCS may be prohibitive and installation can involve major vessel retrofit. Furthermore HFO with high sulphur content is not expected to be as widely available in the future as it is today. It is therefore questionable whether the number of vessels that will be fitted with EGCS by 2020 will be sufficiently large to form a solid market for the continued supply of HSFO.

4. MAIN ENGINE SELECTION

The performance of a vessel depends on many factors, one of the most important and complex being the choice of propulsion system to be installed on the ship. The propulsion system accounts for a large CAPEX percentage of the total shipbuilding cost, as well as a large portion of the annual operating costs in fuel consumption and maintenance. Therefore the selection of the most suitable engine is a crucial factor for shipowners and designers and requires thorough analysis so that the most appropriate propulsion system is selected at an early stage of the vessel design process. The analysis needs to take account of the vessels' design performance requirements, operating profile and recognise that the machinery selection has a major impact on the reliability of a vessel.

For the use of LNG as fuel, there are two distinct engine design concepts available from the major marine large engine manufacturers designed around the Diesel or Otto combustion processes. For the use of LNG the re-gassified LNG needs to be fed into the engine as gas at different pressures, depending on the combustion process. For the Diesel process this is high pressure (approximately 300-350 bar) and for the Otto process this is low pressure (approximately 5-12 bar). This by necessity means there are also significant differences between the Fuel Gas Supply Systems (FGSS) that are required for each engine concept. The different combustion options, run on different cylinder heat cycles and operate with very different characteristics.

The candidate LNG as fuel engines for this study included all possible options of main engines and main engine configurations available on the market, including 2-stroke high pressure and low pressure gas supply engines, and the possibility to adopt a 4-stroke, low gas supply pressure engine.

For the case of the 4-stroke option, the propulsion system selected consisted of two Wärtsilä 31DF engines providing power to a twin input single output gearbox driving a Controllable Pitch Propeller (CPP), delivering an MCR of 8,830 kW at 80.0 rpm. With this configuration, one or two engines can be configured to operate at an optimum efficiency point depending on the vessel speed. A variable speed mode of the engines can also be applied. Auxiliary electrical power (hotel load) can also be produced from these propulsion engines by the use of two shaft generators with frequency converters driven through Power Take Off (PTO) arrangements. With this configuration there is sufficient load capacity to cover the complete range of the vessels' propulsion and generating needs at any operating condition. Therefore the need for additional generator sets to be installed for auxiliary power production can be avoided.

An evaluation system was developed that captured the main trends of the systems involved. The main criteria that the engines were evaluated upon were environmental, performance, financial and operational. The data was collected from publically available sources and some of the findings are depicted in Figure 2.

From the evaluation matrix it was concluded that given the current status of the various technologies, the most suitable propulsion arrangement for the vessel under study was the proposed multi-engine arrangement with 4-stroke DF engines operating on low gas supply pressure and driving through a CPP gearbox. The key factors that led to the decision were the increased safety and redundancy from two engines driving the propeller, high efficiency (due to the high level of technology utilized in the Wärtsilä 31DF design), the clean and environmental friendly operation, and the well proven service record and reliable operation of similar DF engine types.

	2S HP DF	2S LP DF	4S LP DF
THC	☺	😐	😐
CO_2	☺	☺	☺
SOx	☺	☺	☺
NOx	☹	☺	☺
PM	😐	☺	☺
Power density	☺	😐	☺
Efficiency on Gas	☺	☺	☺
Efficiency on Diesel	☺	😐	😐
MN dependency	☺	😐	😐
Transient response	😐	😐	☺
Main engine cost	😐	😐	☺
Gearbox	-	-	☹
Cost of FGSS	☹	☺	☺
Fuel cost	☺	☺	☺
Maintenance cost	☹	☺	😐
Proven service record	😐	😐	☺
Safety / redundancy	😐	😐	☺
Propeller efficiency - LRM	😐	☺	☺
Manoeuvring	😐	😐	☺
Reserve elec. load carrying capacity	😐	😐	☺
Weight compared to standard vessel	😐	😐	☺

Figure 2. Criteria and evaluation scores for three DF propulsion systems

The layout of the propulsion system selected is shown in Figure 3 below.

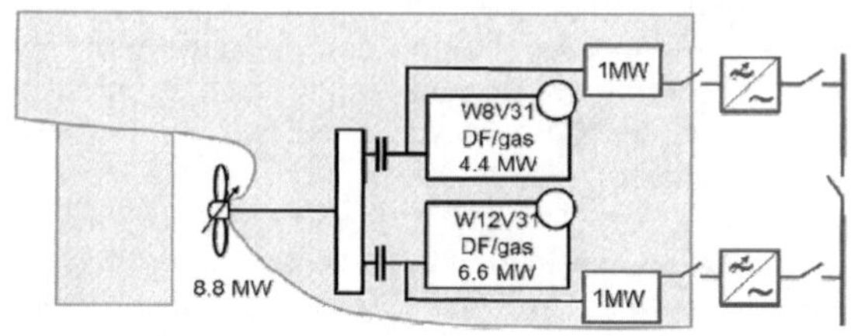

Figure 3. Schematic of propulsion system layout in engine room

Figure 4. Wärtsilä 31DF engine

The Wärtsilä 31DF engine (Figure 4) is a technology enabler for this selection because of the benefits from high efficiency due to the 2-stage turbocharging arrangement, and the NOx Tier III compliance when operating in gas mode which alleviates the need for complex emissions aftertreatment equipment. Furthermore the design of this engine has targeted reducing methane slip by addressing the following:

- Scavenging process, losses during valve overlap: engine has variable valve timing,
- Quenching effects in lean mixture zones: cold areas are avoided by employing high Mean Effective Pressures (MEP),
- Trapped methane in crevice areas: engine has optimized piston and cylinder design,
- Delayed combustion/misfiring: engine has advanced fuel injection technology.

Furthermore, due to the high efficiency the engine has a reduced GHG footprint, in particular in the range of 85% engine load the engine has about 25% less total CO_2 emissions compared to the equivalent diesel version,. These characteristics make it the most suitable candidate for accomplishing the targets of the project.

5. VESSEL DESIGN

Due to the low density of LNG compared to conventional fuel oils, LNG requires much larger storage volumes for the same energy content. Furthermore, the long range operating profile of the vessel required a tank with a volume of 2,500 m^3 which would suffice for 40 sailing days at sea with full speed.

The position of the LNG tank on board could not be selected with the same degree of freedom as with conventional fuel, but instead, needed to comply with the provisions of the IGF code. As mentioned above, the IGF entered into force from 1 January 2017. The IGF Code includes alternative LNG tank location criteria using either deterministic or probabilistic methods; for this study the probabilistic approach based on the calculation of damage probability allowed a reasonably large tank to be accommodated on board within an efficient ship design.

There are several available technologies for the containment of LNG, all of them with very different pros and cons to consider. Major factors that influenced the LNG tank choice for this study were: volumetric efficiency, experience from existing installations, holding pressure and boil-off gas (BoG) rates/management, investment cost, cooling down practices, survey and inspection requirements.

A membrane type tank with holding pressure of 2.0 bar was found to be the optimum solution for the intended LNG volume. This design consisted of corrugated stainless steel with a thickness of 1.2mm which serves as the primary barrier being in contact with the LNG, welded on top of the insulation panel which integrates the secondary barrier between two layers of insulation. The membrane type allowed the maximum utilization of existing space between all containment systems, and at a lower cost and weight.

There are many ways to accommodate the LNG fuel containment system within the ship's hull, the initial approach being to increase the ship's length. This option offered several benefits, such as no loss of cargo space and no penalty on ship's draft, but does increase the ship DWT by 1,500 tonnes. This approach does however give the flexibility to adjust size depending on the particular needs of each owner, no negative impact on ship's stresses and stability, and could be applied for the conversion of young vessels should future conditions (e.g. price of 0.5% sulphur fuel) justify this move.

In many cases, bulk trades are not dependent on vessel draft, since draft restrictions apply at ports where low density cargoes are loaded and the quantity to be loaded is restricted by the total hold volume, and not by the DWT of the ship. Typically heavy cargoes, which require scantling draft, are loaded at ports where there are no draft restrictions. Therefore, the option to increase a ship's depth instead of length remains a possibility. Furthermore, considering the new Panama Canal dimensions, one may also opt to increase the ship's beam. We therefore see that each option requires different extra steel weight and the optimum solution depends on both the specific commercial requirements and the extra cost considerations.

Taking the aforementioned into consideration, the final design selected in this study is depicted in Figure 5, showing the location of the LNG tank of 2500 m^3 capacity and leading to an increase in vessels length of 10 meters. This solution allowed the proposed B.Delta82 LNG vessel to have a new DWT of 83.500t (including an additional 1,500t of cargo), with service speed of 14.5 kn at 7,100 kW.

The overall design details of the vessel include a length of 238m, a breadth of 32.26m, a depth of 20.3m, a draught design of 12.2m, a draught scantling of 14.45m and a DWT scantling of 82,000 tons.

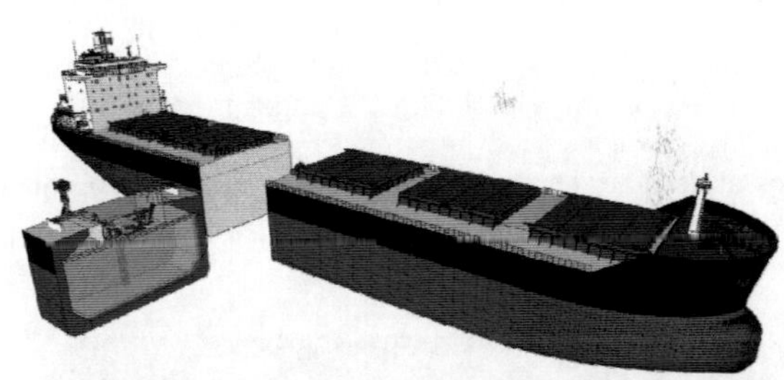

Figure 5. LNG tank on the B.Delta 82 bulk carrier [Gaztransport & Technigaz - GTT]

Due to the heat input to the tank, the liquid evaporates and builds up pressure (natural BoG). While at sea, the total gas consumption/requirement for propulsion and electrical loads is far higher than the natural boil off rate, therefore the BOG can be managed by the demand and the makeup gas is supplied from forced BoG. During stay at port, the natural BoG rate is slightly higher than the quantity required to cover the electrical loads alone. The expected boil off rate under conditions of high air and sea temperature, with the LNG tank being at maximum filling level has been estimated to be approximately 135 kg/h, whereas the electrical load consumption is estimated at approximately 100 kg/h.

Analysis has shown that for various LNG tank filling conditions (42%, 50%, 61%), which are typical for

arrival conditions at ports where a long idle stay is expected such as South America, the gas pressure in the tank would not exceed 1.4 barg after a 40-day period in idle condition, i.e. load only for electrical power.
The current tank is able to hold pressure for more than 15 days without consumption and for more than 30 days with minimum hotel load.

It has also been determined that the LNG filling level at arrival will have a limited influence on the gas pressure increase. The gas pressure increases at a lower rate when the filling level is lower and indeed the absolute amount of BoG decreases with the tank surface area in contact with LNG (wet surface effect), which is dictated by the filling level. The combination of BoG consumption and fuel containment pressure capability therefore maintains LNG tank pressures within acceptable limits for the majority of operational scenarios.

In addition, ageing calculations undertaken during the study have shown that the quality of LNG will remain quite stable during idle periods, and that this phenomenon is not dependant on the LNG containment technology.

The last major influence on the vessel design, from the perspective of the selected propulsion system and the allowance for the LNG tank, involved the effect on the hull lines of the vessel. These were modified accordingly as shown in Figure 6. The specific twin-engine layout allowed for a more streamlined aft part, and further optimisation was achieved by positioning the engines above the gearbox centreline. Optimization and CFD calculations conducted in parallel by both ABS and the vessel designer Deltamarin, succeeded in recovering the extra losses incurred by the extra ship's length.

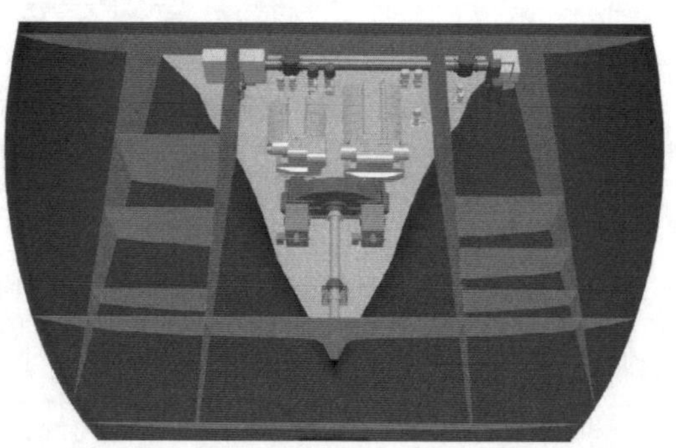

Figure 6. Aft part of the vessel accommodating the selected propulsion system [Deltamarin Oy]

6. CAPEX AND OPEX

The study also compared the machinery options on a CAPEX basis and included comparisons of a typical 2-stroke diesel engine running on liquid fuel and equipped with Tier III NOx technology (e.g. SCR), and the LNG fuelled alternatives. As previously indicated, the study considered two 4-stroke DF engines operating on the Otto combustion cycle in gas mode with low pressure gas injection, a 2-stroke DF engine operating on the Diesel combustion process in both oil and gas modes with high pressure injection, and a 2-stroke DF engine operating on the Otto combustion cycle in gas mode with low pressure injection.

The analysis shown in Figure 7 indicates that although the conventional vessel has the lowest initial cost, the proposed arrangement with two Wärtsilä 31DF engines provides the minimum extra CAPEX among all the considered options for the application of LNG as fuel with the aforementioned technical advantages.

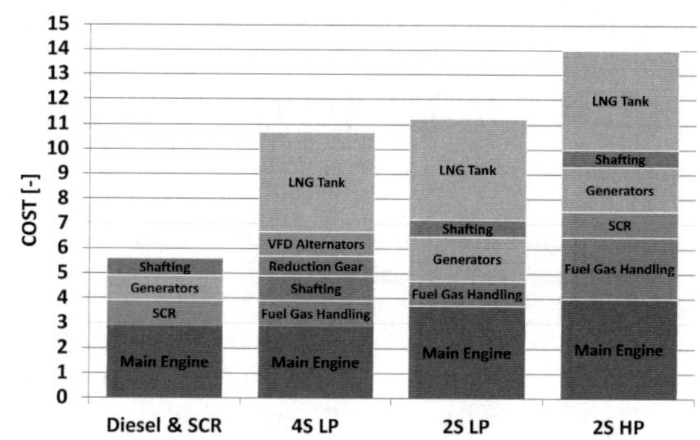

Figure 7. Dimensionless CAPEX comparison for Main Engine and all supporting equipment

6.1 FUEL PRICES

Fuel costs are the biggest factor affecting a ship operator's profitability. At the end of 2016, the purchase price of LNG was approximately 6% lower than that of HFO, on an energy content basis. While fuel and gas prices do fluctuate, the long term trend is expected to favour LNG as the least expensive choice. The fuel efficiency of modern gas-fuelled engines is also higher, with thermal efficiency ranging from 5 to 14% higher than conventional fuelled engines. The business case therefore needs to consider these factors and show that the difference in OPEX, predominantly through the cost of fuel, justifies the decision of adopting the LNG as fuel technologies.

Investigations were therefore carried out in order to determine whether reduced operating expenses (OPEX) could compensate for the increased CAPEX and justify the application of a 4-stroke or a 2-stroke DF engine. The maintenance costs of such engine installations is expected to be comparable to traditional solutions on conventional vessels. Piston overhaul is normally undertaken every five years but due to the cleaner combustion with gas the costs incurred could potentially be lower and the consumption of lubricating oil is also lower. Maintenance costs of the CPP and the reduction gear are not considered significant (for the 4-stroke application). Certain additional maintenance costs related to LNG have not been considered by the study, since similar costs would be accrued by vessels equipped with SCR and SOx EGCS technologies. Long term projections indicate that the price of LNG may range around 9 USD/MMBTU or even lower, including a 3.5 USD/MMBTU premium for delivery onboard (Waterborne Energy, Inc, 2017).

Although it is difficult to predict the price of 0.5% sulphur fuel it is reasonable to assume that as the year 2020 approaches this will rise and become higher than current prices.

In order to examine whether the offset in CAPEX (Figure 7) could be paid back through the different fuel prices, an analysis was conducted comparing the LNG and LSFO prices across several price scenarios. Figure 8 shows that the payback period for the CAPEX could be achieved within 2-3 years, a scenario that clearly justifies the adoption of LNG as fuel technologies.

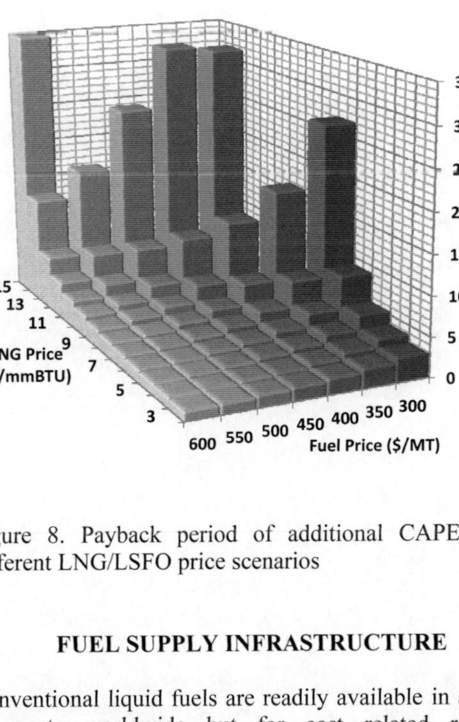

Figure 8. Payback period of additional CAPEX for different LNG/LSFO price scenarios

7. FUEL SUPPLY INFRASTRUCTURE

Conventional liquid fuels are readily available in almost all ports worldwide but for cost related reasons bunkering, in practise, takes place mainly at major ports, the trend is depicted in Figure 9.

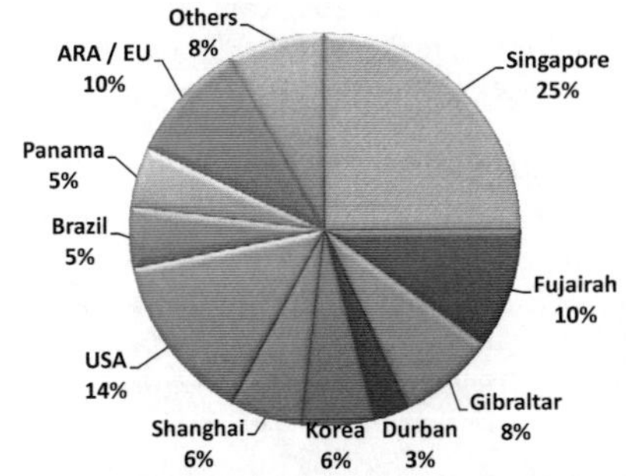

Figure 9. Bunkering uptake of vessels engaged in tramp service

The development of LNG bunkering supply networks in such ports is currently ongoing by major fuel supply companies with initial focus on areas such as the EU, Middle East, Singapore, and the United States.

It is therefore expected that LNG availability in such ports will soon be sufficient to accommodate the expected demand, and support the adoption of LNG as a bunker fuel.

8. CONCLUSIONS

The widespread adoption of LNG as a bunker fuel faces several challenges, including the investment required for the ships propulsion and fuel handling systems, the lack of bunkering facilities and infrastructure, uncertain LNG bunker prices and the necessary development of new international and local safety regulations, particularly with regard to bunkering and shore side LNG supply.

However, the benefits that it offers, for example, the significantly reduced EEDI attributable as a low carbon fuel, the compliance with SOx and NOx Tier III limits without any exhaust emission abatement technologies, and its abundant supply resources (proven reserves 6,900 Trillion cubic feet (US Energy Information Administration, 2016)), can provide the stimulation to overcome these challenges.

The feasibility study described in this paper has demonstrated that through a novel approach to the machinery arrangements, adoption of 'state of the art' prime movers and novel pressurized membrane LNG tanks, it is indeed feasible to use LNG as a bunker fuel on an ocean-going dry bulk carrier. The study shows that the payback period for new construction is viable for the majority of the price scenarios considered. Despite the increased CAPEX compared to conventional ships, the benefits are numerous and demonstrate that the overall investment is expected to pay back in a very short time period with a favourable total cost of ownership.

Therefore, given the potential advantages of LNG as a bunker fuel, and the rate that the LNG infrastructure is developing, it is evident that there is a strong possibility that LNG will capture a significant percentage of the conventional fuel oil market in the near future.

9. REFERENCES

1. SHELL, www.shell.com/marine, April 2017
2. WATERBORNE ENERGY, INC, Aug 2017
3. UNITED STATES ENERGY INFORMATION ADMINISTRATION, www.eia.gov, 2016

10. AUTHORS BIOGRAPHY

Antonis Trakakis holds the current position of Technical Manager at Arista Shipping. He is responsible for the technical management of a young fleet of bulk carriers. As Technical and Environmental Manager of Superfast Ferries he has successfully addressed all issues relevant to operation of ships in sensitive areas like the Baltic Sea.

He has worked with all kinds of internal combustion engines, ranging from gas turbines, to all sizes of 4-stroke, and 2-stroke engines, and one of his special concerns has been the formation and control of emissions, and engine efficiency

Marios Ioannou is a Principal Engineer with the American Bureau of Shipping, in the Corporate Technology department. Prior to that, Dr. Ioannou has worked for many years for a number of engine manufacturers. In his latest role, with Winterthur Gas and Diesel (Wärtsilä) in Switzerland, he was an expert in the engine performance department and was involved in the development of the combustion systems of Diesel and Dual-Fuel Engines, with focus on both, the Diesel and Otto cycles. He was also the Performance Leader during shop tests and Type approval tests at licensee sites and sea trials in China, Korea and Japan.

Dionysios Antonopoulos is a Naval Architect & Marine engineer, currently holding the position of the General Manager, Marine Solutions Sales in Wärtsilä Greece. Mr. Antonopoulos worked for 5 years in Wärtsilä Switzerland, in the department of marketing and newbuilding projects and then appointed as Portfolio Manager for 2-stroke engines, still within Wärtsilä Switzerland, where he cooperated with major Korean and Chinese shipyards for aspects of vessel propulsion and engine development. Mr. Antonopoulos returned to Greece as Technical Manager for 2-stroke engines responsible for Greece and Cyprus and then appointed to the current holding position. Prior to Wärtsilä he worked for a shipping company and he also has seagoing experience.

GTT MARS™, NEW MEMBRANE CONTAINMENT SYSTEM FOR LPG CARRIERS

M Noris and **F Durand,** GTT, France

SUMMARY

This paper describes GTT's new membrane technology which has been designed for the Very Large Gas Carriers. It brings an innovative value proposition to the LPG shipping market making vessels safer, reduces their initial build and operational costs, and increases overall efficiency by allowing greater loads. GTT's technology is a light and compact solution, which enables the design of a more efficient vessel with reduced dimensions and lower fuel consumption, or the optimization of the cargo capacity. This extra cargo could be either monetized: +26% of revenue in the current low freight rate environment (based on a $30/t Baltic Index) for a voyage from Arabian Gulf to India, with two discharges in the East and West coast ; or used for the propulsion of the ship for instance. This will afford more flexibility to switch to LPG when needed, and comply with the new environmental regulation on fuel emissions imposed by the IMO. Besides, the flexibility is increased by the reduction of cooling down time to only 3 hours, which prevent from keeping a heel that can also be monetized. Then, GTT MARS does not suffer any thermal shock. Finally, full scale tests demonstrate the high resistance of the membrane in the event of a collision at sea. In summary, GTT brings four main deliverables to the shipping industry: increased revenue, greater operational flexibility, greener emissions and improved safety.

1. INTRODUCTION

For over 50 years, GTT has been developing and providing efficient solutions for the containment of liquefied gas, both at low temperatures and under cryogenic conditions, for shipping or storage, both onshore and offshore. To handle the cargo, holds need to be coated with cryogenic lining that can withstand the substantial loads. Envelopes, known as membrane, contain the LNG at a temperature up to -163°C, sealing it with a totally impermeable layer between the liquid cargo and the vessel's hull. Creator of the patented concept, GTT is the reference player of the LNG market. The GTT membrane technology allows cost reduction of shipping LNG and loading it, in bulk, into a vessel's holds. The membrane structure fits perfectly to the contours of the vessel's hull, making it possible to optimize storage space and reduce the vessel construction and operation cost. The continuous improvement of GTT solutions allows constant reduction of the boil-off gas rate of the membrane containment systems. Proof of the success of GTT solutions is based on the fact that more than 300 LNG carriers are equipped with membrane technology.

Some of the existing ships, including some of the first LNG carriers, have been operated in the past for LPG and ethylene trade as well (Annabella 35,000 m^3 built in 1975, Ben Franklin 120,000 m^3 built in 1975, Descartes 50,000 m^3 built in 1975). More recently, on board the Shell Prelude Floating Liquefaction platform (FLNG), four tanks are dedicated to LPG condensates, featuring exactly the same technology as the eight tanks dedicated to LNG. Finally, six Very Large Ethane Carriers (VLEC) for Reliance that have been recently delivered by Samsung Heavy Industry also transport LPG. While this demonstrates that the traditional membrane technology is adapted to the carriage of LPG or other cryogenic gases, it is fair to mention that the traditional membrane is overdesigned (design temperature of -163 °C while LPG requires -55°C) and needed to be optimized to the requirements of this very specific LPG market.

Until now, the only solutions available on the market were self-supporting tanks. The objective was to design a dedicated technology tailored to the LPG market for gas carriers, which adds a new value proposal to the transportation of LPG. In an environment with low shipping rates, reducing the total cost of ownership is essential. This is what GTT proposes to offer with this innovative MARS™ technology: cutting construction cost for the shipyards, reducing operational cost and bringing increased safety and greater efficiency to the shipowner.

2. CONTAINMENT SYSTEMS DESIGN

2.1 THE DIFFERENT TYPES OF CONTAINMENT SYSTEMS

Developed by the International Maritime Organization (IMO), the International Code for the Construction and Equipment of Ships Carrying Liquefied Gases in Bulk (IGC Code) governs all gas carriers built after June 1986. According to the IGC code, the entire cargo containment system consists of the following elements: primary barrier (cargo tank), secondary barrier (if fitted), thermal insulation, any intervening spaces, and adjacent structure, if necessary, to support these elements. The IGC Code differentiates between Independent tanks and Integrated tanks. Independent tanks are self-supporting, do not form part of the ship's hull, and are not essential to the hull strength. Membrane tanks, the predominant type of Integrated tank, rely on the mechanical strength of the inner hull for support and consist of two relatively thin continuous membranes to contain the liquefied gas. These are separated by insulated inter-barrier space layers that transmit the cargo load and any internal pressure to the inner hull.

2.2 MARS™ CONTAINMENT SYSTEM

Until now, the only containment system solutions available for the LPG market were using independent tanks, whose steel structures are additional to those of the vessel and which significantly increase the weight of the LPG carrier. With an integrated tank concept specially designed for LPG, MARS™ offers a much lighter containment system and overall vessel lightship weight.

MARS™ containment system is based on a single membrane concept specifically conceived to LPG cargo characteristics: temperature higher than -55°C and density up to 700 kg/m^3. The insulation is fitted on the inner hull of the ship, which acts as the secondary barrier of the tank, and supports the primary barrier. This primary barrier is made of corrugated stainless steel.

2.2 (a) Stainless steel membrane

The primary barrier of the containment system is a stainless steel membrane. Its characteristics are the same as the GTT traditional Mark III primary barrier: a 1.2mm thick, corrugated, 304L stainless steel membrane. MARS™ membrane differentiates from Mark III membrane on the pitch of the corrugation which has been increased from 340mm to 600mm, benefiting from the higher temperatures (-55°C) compared to LNG (-163°C).

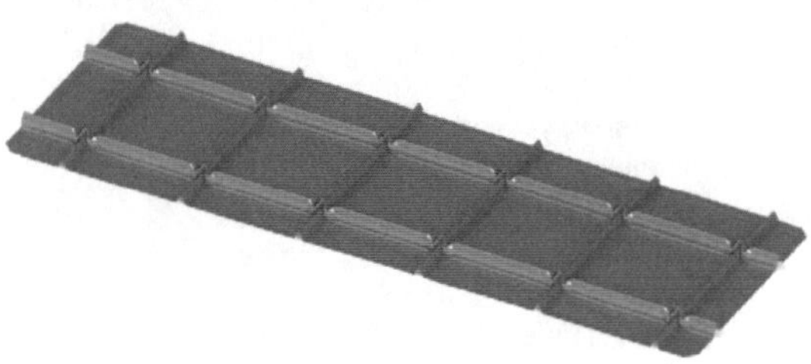

Figure 1: 304L Primary membrane sheet (source GTT)
Insulating plywood boxes

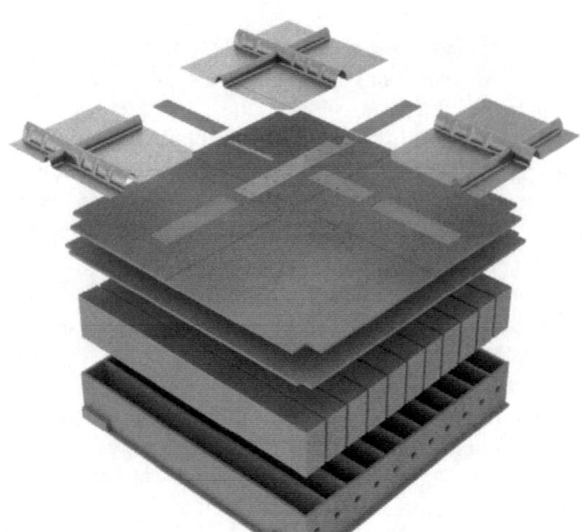

Figure 2: Insulation box 3D exploded view (source GTT)

The membrane is supported by a single layer of insulating material components. The mechanical strength of the insulation is provided by a structure made of plywood. This structure forms a box with several vertical bulkheads that give a high rigidity to the assembly. The thermal performance of the insulation is then ensured by filling the box with glass wool.

The membrane is directly welded on anchoring strips placed on the top of the plywood boxes.

2.2 (b) Mechanical anchoring of the insulation

The containment system is fastened to the inner hull by a purely mechanical system.

Each box is laid on several mastic ropes whose purpose is to compensate the hull deviation and to distribute the loads on the insulation to the hull. The anchoring of the box to the hull is achieved by a coupler system.

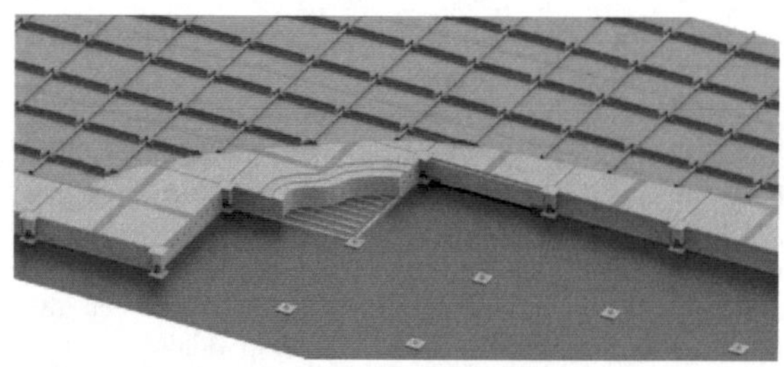

Figure 3: Lay-out in the tank of MARS™ (source GTT)

Figure 4: Coupler system in position in the tank (source GTT)

Each box is maintained at every corner by a setting plate which is pressing the box on the hull thanks to the compression of spring washers. This assembly is attached to the hull by a ball joint connection. Consequently, the anchoring system can freely follow the hull deformations under cargo or ballast pressures.

2.2 (c) Smart corner solutions

MARS™ technology relies on very simple concepts for the special areas of the tank. The solution used for the corners is based on a very simple system of border boxes that allows a highly standardized arrangement.

Figure 5: Corner area arrangement in the tank (source GTT)

The border boxes are made according to the same design philosophy as the flat wall components: plywood boxes, filled with glass wool, that are fastened to the inner hull with mechanical couplers. But these boxes are supporting a steel corner component which is connected to the inner hull by another type of coupler (relying on the same principle as for the flat wall). The membrane is then welded directly to the steel corners. Consequently, when the membrane is pulling on the corners, due to the thermal contraction of the primary barrier or the hull elongation of the ship, the force is not sustained by the insulation but by the inner hull itself.

2.2 (d) Hull acts as the secondary barrier

For cargo temperature over -55°Cn the IGC Code allows to use the hull as the secondary barrier. In this regard, the inner hull of the ship shall be made of low temperature steel (FH grade), as is the case for Type A tanks.

2.3 MARS™ CONTAINMENT SYSTEM FEATURES

2.3 (a) Zero stress concept

The orthogonal corrugation pattern enables the membrane to accommodate any thermal or hull deflection stresses well within the fatigue limit, therefore providing an outstanding lifetime.

2.3 (b) High resistance

The strength of plywood boxes is adjusted to sustain the loads induced by liquid motion in the tank.

2.3 (c) Modular concept

This modular system calls for standard prefabricated components that can accommodate any shape and capacity of tanks and are designed for mass production and simple assembly.

2.3 (d) Safety

The insulation space is continuously flushed with nitrogen gas. The integrity of the membrane is then permanently monitored by detection of hydrocarbon in the nitrogen.

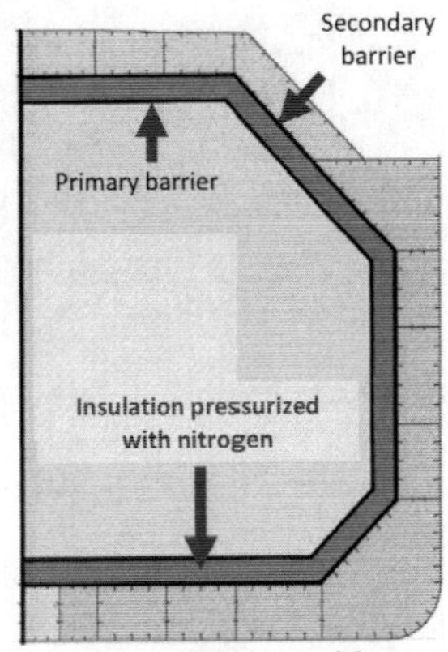

Figure 6: Permanent monitoring with gas detection within insulation space under nitrogen (source GTT)

2.3 (e) Strong resistance to collision

Additionally, the MARS™ stainless steel membrane benefits from a high resistance to collision and grounding due to its flexibility. It can adapt to very large deformation without any loss of tightness. GTT, DNV GL and the Hamburg University of Technology (TUHH) built a large scale mock up, nearly 4 m X 4 m, and tested it experimentally. During the test illustrated in Figures 7 & 8 below, a displacement of 1 m has been measured and small corrugations have been completely unfolded without any crack or loss of integrity.

Figure 7: Mock-up placed in the test rig

Figure 8: Deformation of the mock-up in experiment

The behaviour of the primary membrane under large deformations has been tested separately in cryogenic conditions yielding the same results as the large-scale test.

There have been no incidents or liquid leaks since the membrane has existed.

2.3 (f) Reliability

All materials used in MARS™ are sea-proven in GTT's technologies. Densified wood and glass-wool are easily procured, low cost raw materials, and are available worldwide.

2.4 VALIDATION BY CLASSIFICATION SOCIETIES

Classification Societies have been able to fully review each step of the development of MARS™. The four main classifications societies Lloyds Register, DNV GL, ABS and BV granted their Approval In Principle (AIP). In April 2017, after a deep review of the technology and of all support calculations notes and tests, Lloyds Register granted their General Approval for all the critical points of design of the containment system. This is a very important milestone for the development of the technology as it demonstrates MARS™ meets the gas and maritime industry requirements and is ready to be ordered.

3. IMPACT ON SHIP's DESIGN

3.1 LOWER LIGHTSHIP WEIGHT

Because it is fully integrated, MARS™ makes it possible to turn this reduction of weight into a cargo capacity increase. The solution also adapts to any shape and size of ship.

To illustrate this benefit, GTT has performed a basic design study of a typical 84,000m^3 by starting from an actual design of Very Large Gas Carrier (VLGC) equipped with Type A tanks.

While keeping the same overall dimensions and the same propulsion, the ship has been modified to replace the Type A tanks with MARS™ tank type technology. As a result, the single hull has been replaced by a double hull in order to have a flat surface inside the hold to fit the membrane.

In the end, considering that the 4 Type A tanks (≈5,000 tons) were removed, that the hull has been modified (around 2,700 additional tons of steel for double hull) and fitted with MARS™ tanks (≈1,300 tons), the MARS™ solution provides a weight reduction of 1,000 tons on the 84,000 m^3 design.
Therefore, if a 84,000 m^3 vessel equipped with a self-supporting tank weighs 19,000 tons, the same vessel equipped with GTT's technology would weigh only 18,000 tons.
That reduction in weight translates immediately into increased cargo capacity: removing 1,000 tons of steel makes room for 2,000 m^3 more cargo.

3.2 TANK SHAPES FLEXIBILITY

MARS™ provides more flexibility in terms of tank shapes providing more options to ship's designer to optimize the vessel performances.

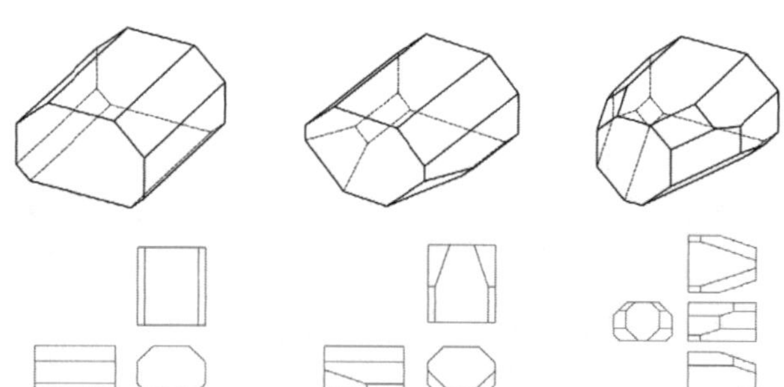

Figure 9: Three different types of shapes possible (source GTT)

3.3 GTT MARS™ 91,000 m^3 SHIP DESIGN

GTT's 91k design is a vessel that has the same dimensions as a typical 84k Type A ship:

- Length O.A.: 228.6m
- Breadth: 36.6m
- Design draft: 11.8m
- Service speed: 16.5Kn

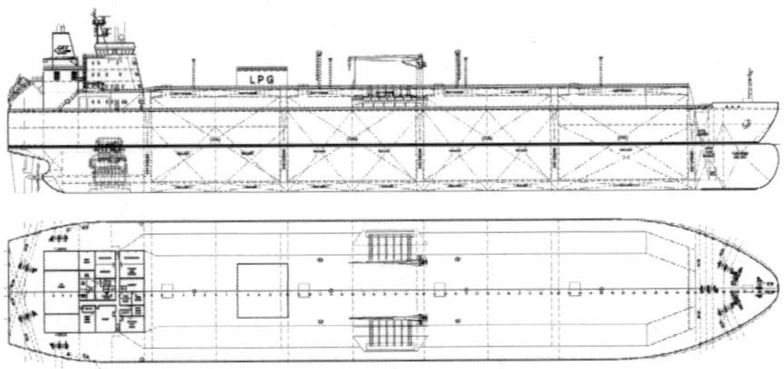

Figure 10: GTT Membrane 91k ship design (source GTT)

Cargo capacity is then adjusted on liquid density: 91,700 m^3 of propane (density=0.58) or 86,000 m^3 of propene (density=0.62) can be loaded, keeping the draft constant at 11.8m. This means an additional capacity up to 7,000 m^3.

Another philosophy of design could fit with the advantages given by the membrane: a 84,000 m^3 vessels, with reduced dimensions, and so less fuel consumption.

3.4 IMPACT ON SHIP'S CONSTRUCTION

The industrial scheme is being developed with selected shipyards. It is expected that the construction time for a vessel equipped with MARS™ will be similar to vessels equipped with other containment systems.

4. IMPACT ON SHIP'S OPERATION

MARS™ is a light and compact solution, which enables the design of a more efficient vessel with reduced dimensions, lower fuel consumption, or the optimization of the cargo capacity. This extra cargo could be either monetized or used for the propulsion of the ship for instance. This will provide more flexibility to switch to LPG when needed, and comply with the new environmental regulation on fuel emissions imposed by the IMO.

4.1 HEAT CAPACITY OF THE CONTAINMENT SYSTEM

MARS™ is mainly composed of plywood (with remaining steel and glass wool) while self- supporting tank technology is exclusively composed of steel (and some foam). Additionally, MARS™ tank is about 1/3 of the weight of an equivalent self-supporting tank. Those two effects together result in a heat capacity about three times lower for a MARS™ tank compared to Type A tanks. It takes one third as much energy and quantity of LPG to cool down the tank. Cooling down operation time is then significantly reduced: 3 hours with GTT membrane technology compared to 20 hours for a self-supporting tank.

4.2 RESISTANCE TO THERMAL SHOCK

Because of the ability of the membrane to absorb thermal elongation, due to the corrugation patterns, MARS™ is not sensitive to thermal shock and loading can be done without the need to pre-cool the tank.

The above features have significant impact on the OPEX (OPErating EXpenditures) and flexibility in operation.

4.3 THE FUTURE OF LPG CARRIER OPERATION

GTT's technology is a light and compact solution, which enables the design of a more efficient vessel with reduced dimensions and lower fuel consumption, or the optimization of the cargo capacity.

4.3 (a) Additional revenue

This extra cargo could be monetized and represents an additional +26% of revenue in the current low freight rate environment (based on a $30/t Baltic Index) for a voyage from Arabian Gulf to India, with two discharges in the East and West coast.

4.3 (b) LPG as a fuel

This extra cargo could also be used for the propulsion of the ship. This will afford more flexibility to switch to LPG when needed, and comply with the new environmental regulation on fuel emissions imposed by the IMO in 2020.

4.3 (c) Other operations

This extra cargo could also give greater flexibility to the operators, for many different operations: STS is one of them.

5. CONCLUSION

Simpler than membrane technologies for LNG carriers, MARS™ technology is based on the state-of-the-art GTT membrane. Its safety and quality track record, with reduced cooling-down duration, low heat capacity, lower lightship weight and high resistance to collision offer significant advantages to shipowners.

MARS™ brings increased efficiency, greater operational flexibility and improved safety to the LPG market.

6. REFERENCES

1. ELHERS, G., '*Experimental and numerical analysis of a membrane cargo containment*

system for Liquified Natural Gas', Ships and Offshore Structures, 2016

7. AUTHORS BIOGRAPHY

Mélodie Noris holds the current position of Business Development Manager at GTT. She is responsible for the commercial development of GTT MARS. Her previous experience includes oil market analysis in the commodity market research team for SGCIB New York, strategy consultancy in the oil & gas field, and market analysis in the Strategy and Marketing Department of GTT.

François Durand holds the current position of Project Leader at GTT. He is responsible for the development of MARS containment system. His previous experience includes the management of LNGC projects and the study of containment solutions dedicated to ethane market.

THE USE OF WASTE HEAT RECOVERY TO IMPROVE THE EFFICIENCY OF DUAL FUEL DIESEL ELECTRIC LNG CARRIERS

E E Attah, Maritime Centre of Excellence, NLNG Ship Management Limited, Nigeria; and Department of Mechanical Engineering, University College London, UK, **M Ebong**, Maritime Centre of Excellence, NLNG Ship Management Limited, Nigeria and **R Bucknall**, Department of Mechanical Engineering, University College London, UK

SUMMARY

New Energy Efficiency Design Index (EEDI) regulations and the development of 2-stroke natural gas (NG) burning diesel-engines have put pressure on the Dual Fuel Diesel Engine (DFDE) propulsion system, which is gradually displacing the low efficient steam turbine propulsion system, as the preferred choice for future Liquid Natural Gas carriers (LNGCs). This has led to innovations and developments to improve the efficiency of DFDE LNGCs. One such development is the adoption of the exhaust gas waste heat recovery system (EGWHRS) to improve efficiency. This paper investigates the practical application of an EGWHRS and how it impacts on the efficiency of DFDE vessels. From this study, it has been found that the use the EGWHRS gives an increase in efficiency of 6.6% during practical sea going operations resulting in considerable financial savings over a one year period. The results also indicated that the efficiency improvements based on the use of the EGWHRS ensures that current designs fitted with this system will meet the EEDI baseline regulations mandated by the IMO from 2025 onwards.

ABBREVIATION

BOG	Boil Off Gas
BOR	Boil Off Rate
C_f	Carbon Factor
CO_2	Carbon Dioxide
DDD	Direct Drive Diesel
DDD-GI	Direct Drive Diesel Gas- Injection
DFDE	Dual Fuel Diesel Electric
DWT	Deadweight (in Tonnes)
EEDI	Energy Efficiency Design Index (gCO_2/tNM)
EGWHRS	Exhaust Gas Waste Heat Recovery System
FBO	Forced Boil Off
GCU	Gas Combustion Unit
GHG	Green House Gases
HFO	Heavy Fuel Oil
ICE	Internal Combustion Engine
IMO	International Maritime Organisation
LNGC	Liquefied Natural Gas Carrier
MARPOL	Marine Pollution
ME-GI	M-Type Electronically controlled Gas Injection
MEPC	Marine Environment Protection Committee
OPEX	Operating Expenditure
SFC	Specific Fuel Consumption (g/KWh)
SSDE-GI	Slow Speed Diesel Engine- Gas Injection
SSDR	Slow Speed Diesel with Re-liquefaction plant
STPS	Steam Turbine Propulsion System
WHRS	Waste Heat Recovery System

1. INTRODUCTION

The use of natural gas (NG) has increased dramatically over the past two decades. This is mainly as a result of supply abundance, low cost, and that NG produces less atmospheric emissions, in particular Carbon Dioxide (CO_2), than the other established fossil fuels. This has led to NG becoming the dominant fuel for electricity generation and heating in many countries.

To meet ongoing rise in demand, Natural Gas (NG) is increasingly being stored and transported in liquid form, Liquefied Natural Gas (LNG), by cooling to -160°C it occupies 600 times less volume thereby making it easier and cheaper to transport to market by use of specially designed cryogenic vessels - LNG Carriers (LNGC).

Use of LNGCs is a cost effective way of transporting NG over long distances where long transmission pipelines do not exist. However transport must occur between specially designed LNG terminals. At the loading port NG is liquified and loaded into the LNGCs whilst at the unloading port the LNG is re-gasified, after which it is typically distributed by a national grid pipeline to consumers (US Department of Energy, 2017).

As at September 2017 there are 495 LNGCs in operation (Clarksons, 2014a) which range from the smaller vessels used for coastal operations (Length: 80m Capacity: 2500m^3) to the large QMax type (Length: 345m, Capacity: 267,000m^3), although the majority of LNGCs are in the Length: 275m – 285m, Capacity 135,000 m^3 - 175,000 m^3 range. These LNGCs utilise three different propulsion systems with steam turbine propulsion (STPS) being widely used, whilst the remainder are propelled by either diesel electric (DFDE) or the slow speed 2-stroke diesel engine with the latter usually with a re-liquefaction plant.

LNG is transported at -160°C and at near atmospheric pressure in unpressurised highly insulated tanks. Due to imperfect insulation and sloshing in the tanks on passage, the LNG cargo boils off slowly. It is of great importance that LNGCs have the means to handle the boil off gas (BOG) on-board the vessel. NG is a known greenhouse gas so cannot be vented to atmosphere and the loss of BOG would potentially negate the economic case. It is this need to handle BOG and the fact that the use of NG to fuel boilers is relatively simple that has influenced LNGCs to employ the SPTS over other propulsion systems since LNGCs were first introduced in October 1958.

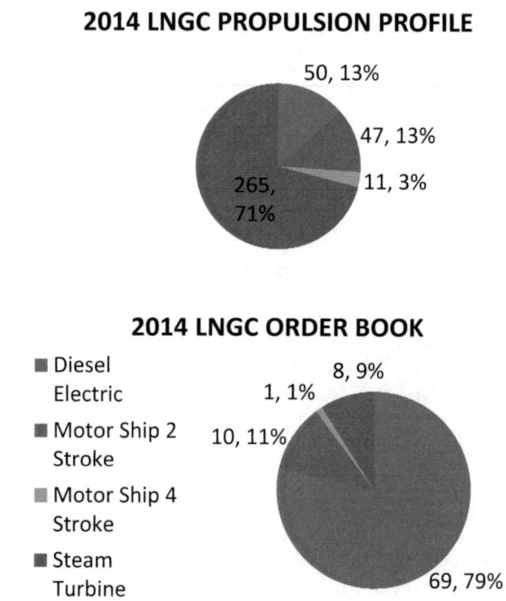

Figure 1: 2014 LNG Fleet and Order book (Clarksons, 2014b)

SPTS has proved extremely reliable in LNGCs but is a less efficient propulsion system when compared to other propulsion types used by other merchant ships. This is due to the inherent constraints of the Rankine Cycle. Initially, there was little motivation to develop more efficient types of propulsion system for LNGCs since the state of the LNG tank insulation was such that on loaded voyages, the natural boil off flow from the cargo was sufficient to satisfy the fuel requirement for propulsion. Even if more efficient gas burning propulsion plants had been available at that time, such designs could not have been used without providing an alternative solution for dealing with the excess boil off gas (BOG) (Wayne & Hogson, 2006).

In 2014, DFDE propulsion was the main alternative choice for modern LNGCs to replace STPS in new builds as shown in Figure 1. The higher thermal efficiency of the DFDE offered significant financial savings when operating the vessel (OPEX) over a LNGC's life cycle compared to STPS. However, Greenhouse Gas Emissions (GHG) reduction regulations such as the Energy Efficiency Design Index (EEDI) and newer competing LNG propulsion technologies including the gas-injection slow-speed 2-stroke engines (SSDE-GI) have highlighted a need to improve the efficiency of the current DFDE propulsion if it is to remain competitive. This is particularly important as analysis of the different propulsion options plant efficiencies highlights that the slow speed 2-stroke engine has the highest efficiency of the options available to LNG shipping as shown in Table 1.

More recently, improvement in insulation technology in LNGCs has resulted in significantly lower LNG cargo boil off rates leading to there now being insufficient natural gas boil off to fuel the LNGC. Therefore there is a requirement for the use of forced boil off gas or use of an alternative fuel such IFO180 or HFO360 to supplement the BOG shortfall (Wayne & Hogson, 2006).

Consequently, there has been renewed interest in using more fuel efficient propulsion systems and/or re-liquefaction plant (converting BOG back to LNG) to gain economic advantage over STPS. Furthermore, over the same period the International Maritime Organisation (IMO) has begun to introduce tougher regulations governing the emissions in exhaust gases from ships including Oxides of Nitrogen (NOx) and Carbon Dioxide (CO_2). The result has been a shift from the STPS towards the DFDE and Slow Speed Diesel Engine (SSDE). When analysing statistical data from Clarkson's World Fleet Register (Clarksons, 2014b) it is found that from the 1960s through to the 1990s, STPS was employed by nearly all LNG carriers. By the 2000s, the proportion of STPS installations had fallen to 69% and in the current decade this has fallen to 21%. When considering the future order book STPS is expected to fall to 12% (Clarksons, 2014b). However the direction of travel is uncertain due to rapid development in propulsion technologies.

The following information supports Table 1:

- For the steam turbine solution, the fuel which in this case includes BOG, is used by the boilers, it has 89% efficiency, and goes through the steam cycle which itself has a 35% efficiency, then to the shafting system which is about 99% efficient, giving an overall efficiency of approximately 30%
- For the DFDE, fuel- which in this case also includes the BOG, is fed to dual fuel diesel engines which are 48% efficient, from there the energy goes through a mechanical/electrical conversion of which has a 98% efficiency, from there it goes through an electrical/mechanical electrical conversion for the propulsive motors which is also 98% efficient, and through the gear box and shafting both of which have efficiencies of 98% and 99% respectively. This brings the total propulsive power efficiency to approximately 43%.
- For the 2-stoke diesel, the fuel- in this case just the HFO, is put into diesel engines which are 49% efficient, from there it is connected to the shafting system which is about 98% efficient. The result of which is an overall efficiency of about 48%

Table 1: Average Efficiencies of the LNGC Propulsion Options. Source (Adrianos, 2006)

	Steam		**DFDE**		**2-Stroke Diesel**	
Plant Efficiencies;	Fuel/BOG	100%	Fuel/BOG	100%	Fuel	100%
	Boilers	89%	DF Engines	48%	Engines	49%
	St'm Cycle	35%	Conversion	98%	Shafting	98%
	Shafting	99%	Motors	98%		
			Gearbox	98%		
			Shafting	99%		
	Propulsion Eff'cy: 30%		**Propulsion Eff'cy: 43%**		**Propulsion Eff'cy: 48%**	
	Fuel BOG	100%	Fuel/BOG	100%	Fuel	100%
	Boilers	89%	DF Engines	48%	Aux Engines	45%
	St'm Cycle	30%	Alternators	97%	Conversion	96%
	Conversion	96%				
	Electric Pwr Eff'y: 26%		**Electric Pwr Eff'y: 47%**		**Electric Pwr Eff'y: 43%**	

The recent development of the SSDE-GI has led to a paradigm shift in propulsion choice for LNGCs over the last three years. Analysing of the order book in 2014 showed that the DFDE was the propulsion system of choice but by 2017 its position had been displaced by the SSDE-GI (Clarksons, 2017).

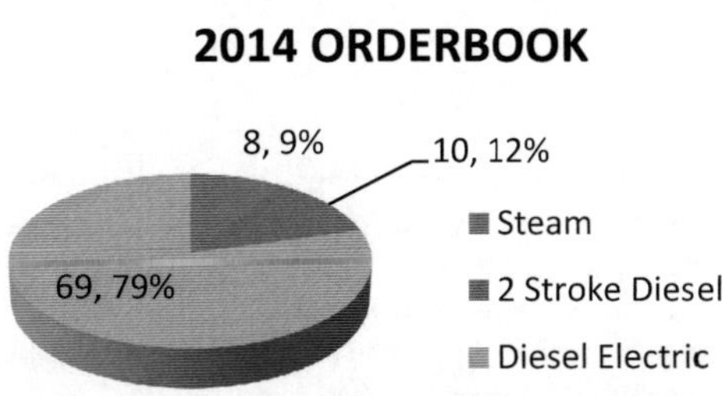

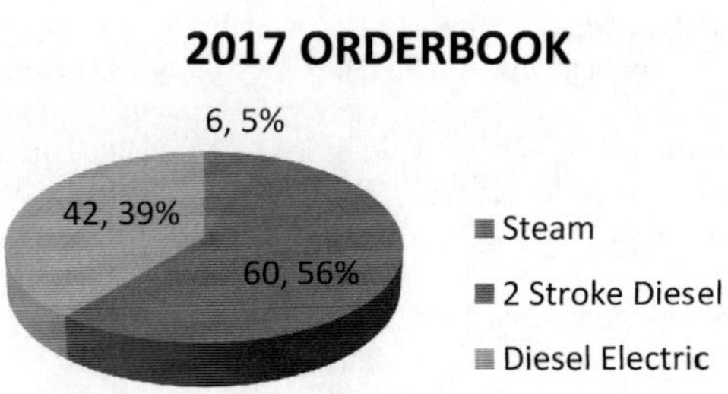

Figure 2: 2014/2017 LNG fleet Order book (Clarksons, 2017)

The development of SSDE-GI technology has led to the need for improvement in DFDE technology if it is to maintain its competitiveness. A range of possibilities exist with improvements in electrical machine technology, more efficient power electronics, use of energy storage to optimise diesel engine loading, etc. and much work has been published. This paper however focuses is the development of an exhaust gas waste heat recovery system (EGWHR) as one of the ways to improve the efficiency of LNGCs using DFDE. Three major aspects are covered: 1) The EGWHRS design 2) Trial Run of the EGWHR in sea-going operations 3) A discussion of results together with a discussion on the implications for future designs. The paper is written from both an academic and practical view point and contains observations from on-board a modern seagoing DFDE LNGC.

2. DESCRIPTION OF THE DFDE PROPULSION

Figure 3 shows a schematic of a typical DFDE propulsion system which consists of a number of, (typically four), medium-speed diesel engines modified to burn BOG in addition to MDO/IFO180/HFO360. The diesel-engines operate at near constant speed and drive alternators that generate three-phase electric power typically at 6.6 kV and 60 Hz supplied to a common busbar system (usually four diesel generators divided across a two busbar arrangement with an interconnector breaker to provide for redundancy). The provision of multiple diesel generators also provides flexibility of operation. This is termed the power station concept because the generated electrical power supplies both the propulsion system and the ship's services (Wayne & Hogson, 2006). The duel-fuel diesel engines operate mainly on NG using a small amount of diesel for ignition purposes. The engines cannot burn the two fuels in variable proportions at the same time in current designs. This is actually a disadvantage compared to STPS which can burn both fuels in various proportions simultaneously. The generated electricity is at constant voltage and frequency, the bulk of which is converted by a power converter, consisting of power electronic devices, for the purposes of controlling the speed of the propulsion motor and fixed pitch propeller whilst a lesser amount is used for ship's services. This means that BOG can be used to provide for all the ship's energy needs.

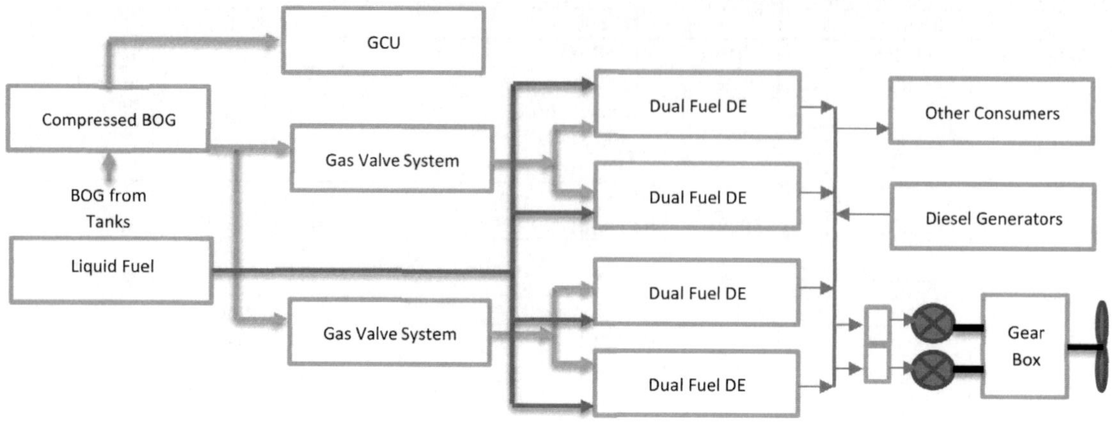

Figure 3: Schematic of the DFDE. Source: (Chang, *et al*, 2008)

The requirement of handling BOG in instances when the main propulsion system is not in use and there is insufficient service load e.g. at anchor or drifting, necessitates the need for a Gas Combustion Unit (GCU), the capacity of which is usually matched to equal to the total BOG capacity on a typical laden journey.

In terms of efficiency, the DFDE technology offers an improvement on the previous LNGC propulsion options, namely STPS and Direct Drive Diesel with re-liquefaction plant (DDD+R). This is because, the steam system offers a thermal efficiency of about 30-35% compared to DFDE of approximately 43%. For the direct drive option, despite its higher engine efficiency (48%), the requirement for a re-liquefaction plants and its large power demand reduces its overall efficiency to about 42% (Adrianos, 2006).

Two current propulsion developments have changed the balance. The first being the SSDE-GI propulsion option which is a Direct Drive Diesel capable of utilising the boil off gas as a fuel, thereby negating the need for a re-liquefaction plant. This means that the SSDE-GI offers a higher propulsive efficiency than the DFDE and is likely to become more attractive as the propulsion option for LNGCs as confirmed by the 2017 orderbook. The second development is the Energy Efficiency Design Index (EEDI) which mandates an improvement in technological options from current levels by at least 30% by 2025. Both these developments have made it imperative to seek improvements in the DFDE technology hence the need to consider an Exhaust Gas Waste Heat Recovery System (EGWHR) as one option to improve DFDE efficiency.

3. DESIGN CONCEPT

It has been estimated that between 48 to 51% of the total heat energy of ICEs is lost as waste heat; therefore the EGWHRS is considered an effective tool to capture and reclaim this heat energy and increase the overall efficiency of the plant. Figure 4 highlights the losses from the DFDE as well as highlighting possible recovery options.

4-Stroke engines tend to have higher exhaust temperature than 2-stroke engines (Olsen, 2016), hence the DFDE has a higher potential for exhaust gas waste heat recovery. An EGWHR system generates steam which can be used for various applications, including electricity generation, contributing to increased efficiency. One study estimated the overall service load consumption i.e. generated power and energy used for auxiliary systems, could be up to 3MW (Olsen, 2016).

The design here is for five exhaust gas economizers to be installed in the exhaust gas path where the waste heat of the DFDE is used for generating steam to drive a turbo generator. The steam is also available for other steam consumers on board the ship. The economizer itself is comprised of a forced evaporator and a superheater, all connected to a common steam drum and directed to the steam turbine. The steam turbine is brought online when all the diesels units are running on at least 80% load so that there is sufficient steam available for the turbine to function reliably. Figure 5 shows a simple design configuration of the EGWHRS.

The Full Schematic of the configuration is shown in Figure 6.

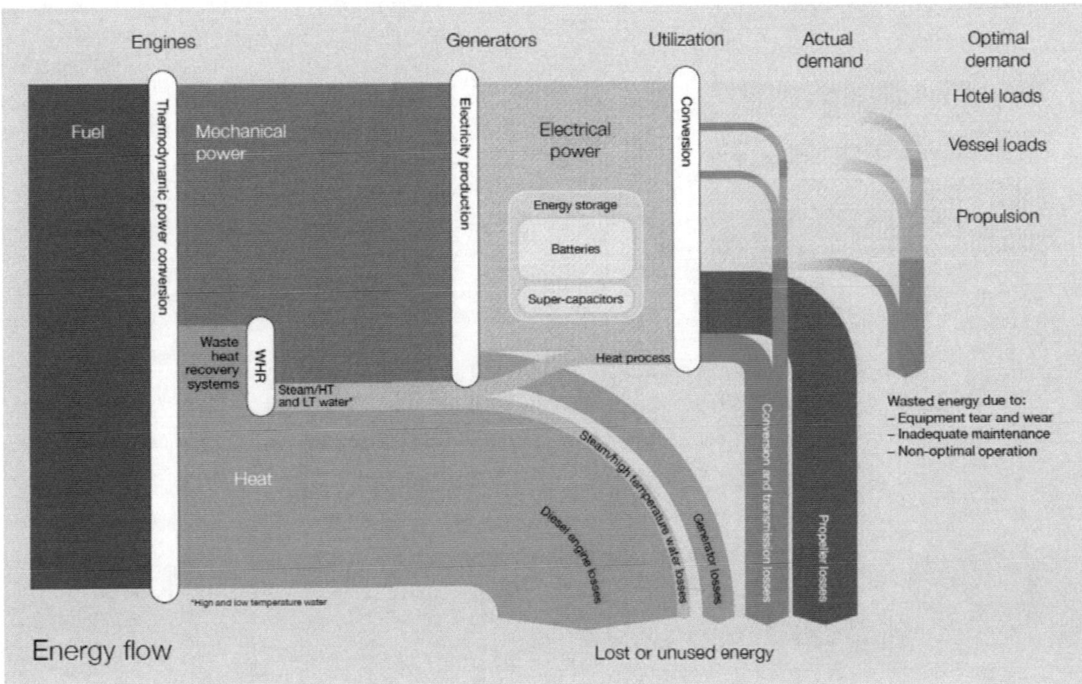

Figure 4: 2014/2017 Sankey Diagram of energy consumption in a DFE (MAN Diesel & Turbo, 2016)

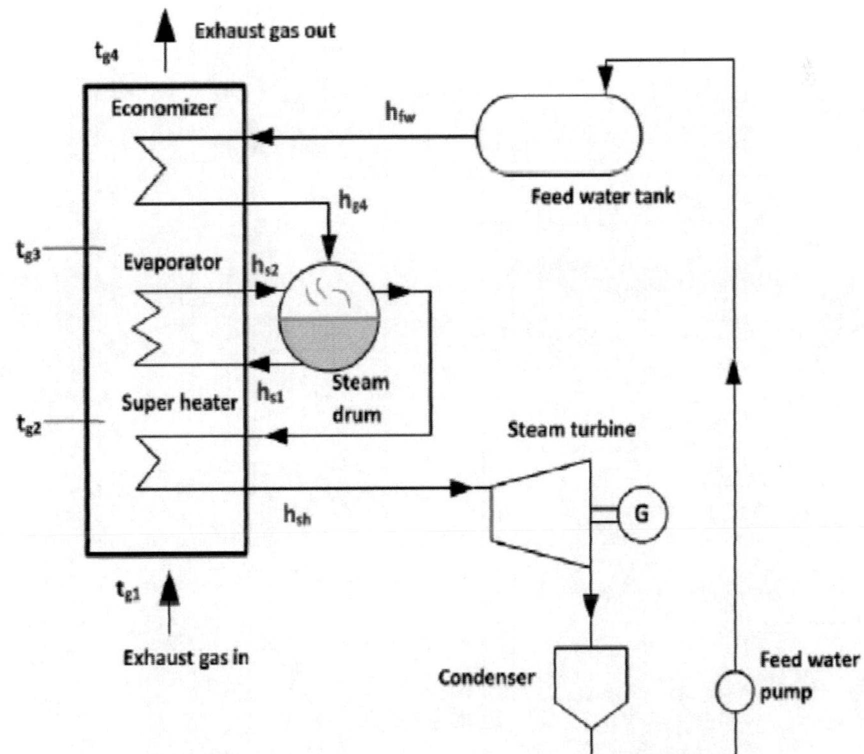

Figure 5: Configuration of an EGWHRS driving to Turbine Generator. Source: (Senary, *et al*, 2016)

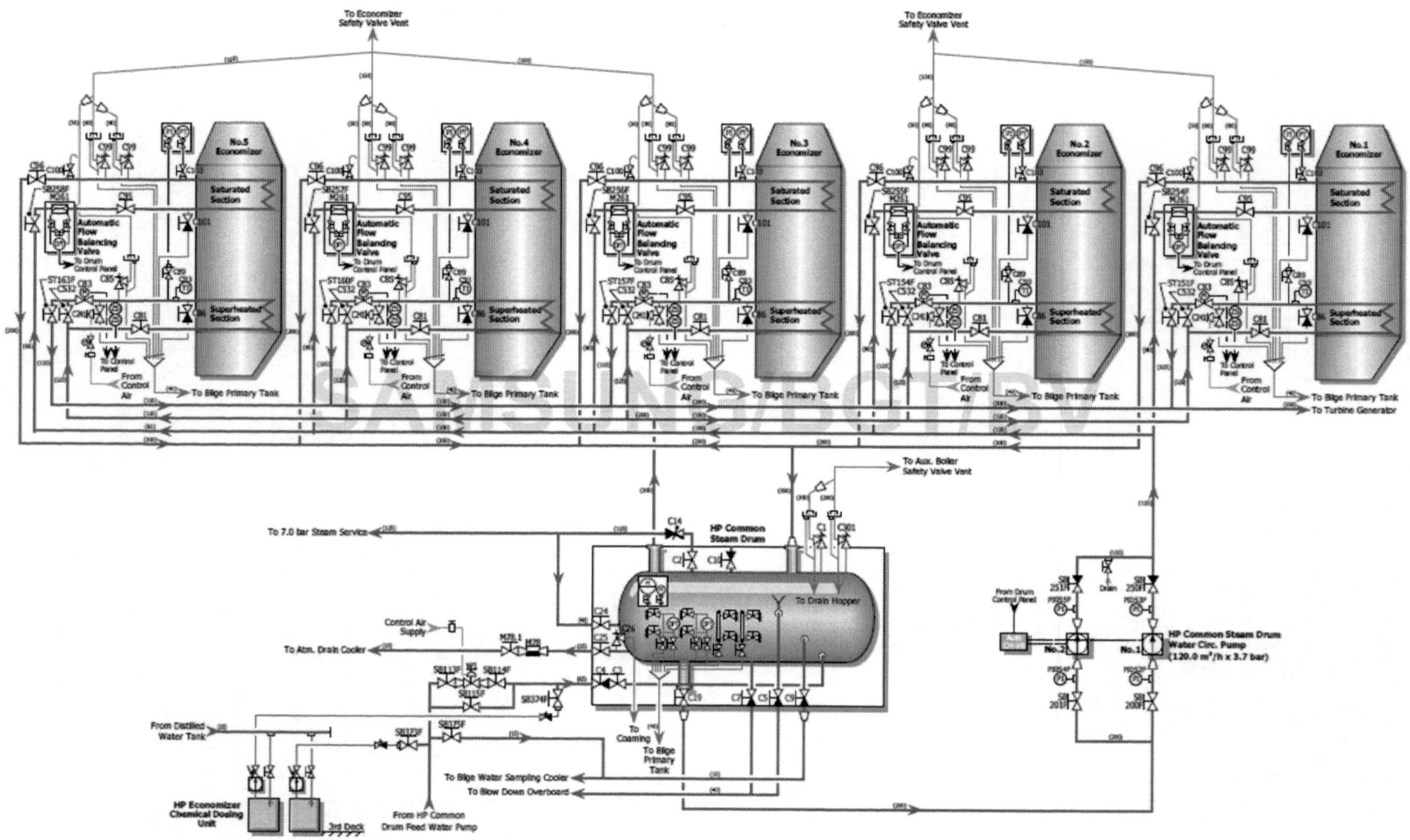

Figure 6: Full Configuration of the EGWHRS used on Trial Ship. Source: (US Department of Energy, 2017)

Table 2: Trial Vessel Particulars. Source (SHI, 2015)

Principal Ship Particulars		**Exhaust Gas Economiser**	
Shipbuilder	Samsung Heavy Ind.	Maker	Alfa Laval
Type of Ship & Cargo	DF-Electric LNG Carrier	Type	Aalborg XW-TG/Q5
Cargo Tanks	4GTT Mark III	No. of Sets	5
Cargo Capacity	174904.2m^3	Capacity	400 kg/h each
Length Overall	293.193m	Design Engine Load	80% MCR
LBP	280.0m	Ex. Gas Mass Flow	34560 kg/h (Gas)
Design Draft	11.65m	Ex. Gas inlet Temp	404°C (Gas)
Deadweight		Ex. Gas outlet Temp.	187°C (Gas)
Main Generator Engine		**Diesel Engine Generators**	
Maker	Wartsila	Maker	GE Energy
Type	8L50DF	Number of Sets	5
Number of Sets	5	Output	8367KVA (7530KW)
Output	7800KW	Voltage	AC 6600V/732A
Speed	514rpm	Frequency/Speed	60 HZ/514rpm

4. TRIAL RUN

The details of the ship and relevant equipment where the trail run was held are shown in Table 2 (Olsen, 2016).

The test involved running the above vessel in two modes in order to ascertain the effectiveness of the EGWHRS and determine the reductions in fuel consumption and emissions when the equipment is in use:

- Mode 1: Vessel is fully laden and en-route at service speed. The test is carried out for two hours between 09:00 hours and 11:00 Hours with four of the DFDE engines online at over 80% load. The Turbo Generator is not however in use and the parameters are monitored and recorded.
- Mode 2: Upon completion of Mode 1 the EGWHRS is brought online. The test is then repeated for two hours between the hours of 12:00Hrs and 14:00Hrs with the DFDE engines under the same load conditions as in Mode 1. The Turbo Generator is however in service and electricity generated is fed to the main switchboard.

The next stage involves estimating the running hours of the TG during actual operations so as to ascertain the effectiveness of the EGWHRS during the actual voyage operations of the vessel. This trial was conducted on the four vessels where the EGWHRS was installed for a month of operation and included laden and ballast passages.

5. RESULTS

The results from Mode 1 and Mode 2 are represented in Table 3 and Table 4 (Appendices). The items recorded included test duration, ship condition, sea state, engine consumption, and auxiliary load data.

In Mode 2, shown in the Table 4, the same data (test duration, ship condition, sea state, engine consumption, and auxiliary load) recorded in Mode 1 were also recorded and monitored for conformity to ensure accuracy of the analysis being carried out.

The second case study involves the monitoring of the Turbo Generator's use across the other four vessels fitted with the EGWHRS over a one month period and recording the data to be used in the analysis. The data recorded includes date, average TG power, running hours, and average speed. The full results are summarized in Table 5 (Appendices).

6. ANALYSIS OF RESULTS

From the results the following calculations were made in order to compare the fuel savings for the different modes of operation:

Table 6: Test 1 Results Summary

Mode 1 (Without Turbo Generator)		**Mode 2 (With Turbo Generator)**	
Total Gas consumption during test(T)	8.6456	Total Gas consumption during test(T)	8.0722
Total Gas consumption during test (m^3)	19.273	Total Gas consumption during test (m^3)	17.995
Equivalent Gas combustion 24Hrs (m^3)	231.27	Equivalent Gas combustion 24Hrs (m^3)	215.94
Gas Density: 0.448588 tons/m^3 GCV (Mass) : 54.8472 MJ/Kg			

Table 7: Updated Efficiencies of LNGCs

	Steam		**DFDE**		**2-Stroke Diesel**	
Plant Efficiencies;	Fuel/BOG	100%	Fuel/BOG	100%	Fuel	100%
	Boilers	89%	DF Engines	54%	Engines	49%
	St'm Cycle	35%	Conversion	98%	Shafting	98%
	Shafting	99%	Motors	97%		
			Gearbox	98%		
			Shafting	99%		
	Propulsion Eff'cy: 30%		**Propulsion Eff'cy: 48%**		**Propulsion Eff'cy: 48%**	
	Fuel BOG	100%	Fuel/BOG	100%	Fuel	100%
	Boilers	89%	DF Engines	54%	Aux Engines	45%
	St'm Cycle	30%	Alternators	97%	Conversion	96%
	Conversion	96%				
	Electric Pwr Eff'y: 26%		**Electric Pwr Eff'y: 50%**		**Electric Pwr Eff'y: 43%**	

In Mode 1, the following calculations have been applied to determine the gas consumption in 24 hours using steady state conditions:

$$Gas\ Consumption\ During\ Test = 8.645 Tonnes$$

$$Gas\ Consumption\ during\ Test = 8.6456 \div 0.448588 = 19.273 m^3\ (Gas\ Density\ 0.448588 tons/m^3)$$

$$Equivalent\ Gas\ Consumption\ in\ 24\ Hours = 19.273 m^3 \times 12 = 231.75 m^3$$

In Mode 2, the following calculations have been applied to determine the gas consumption in 24 hours using steady state conditions

$$Gas\ Consumption\ During\ Test = 8.0722 Tonnes$$

$$Gas\ Consumption\ during\ Test = 8.0722 \div 0.448588 = 17.995 m^3\ (Gas\ Density\ 0.448588 tons/m^3)$$

$$Equivalent\ Gas\ Consumption\ in\ 24\ Hours = 17.995 m^3 \times 12 = 215.94 m^3$$

From the above the following can be ascertained:

$$Gas\ consumed\ in\ 24\ Hours\ without\ WHRS = 231.275 m^3$$

$$Gas\ consumed\ in\ 24\ Hours\ with\ WHRS = 215.9362 m^3$$

$$Fuel\ Saved\ in\ 24\ Hours = 15.3388 m^3$$

$$Percentage\ improvement\ in\ Efficiency = {}^{15.3388}/_{231.275} = 6.6\%$$

6.1 EFFICIENCY IMPLICATIONS OF RESULTS

In terms of thermal efficiency, a 6.6% increase in efficiency will amount to an improvement in the dual fuel engine efficiency from 48% - 54%, leading to an overall propulsion efficiency to 48% making it comparative to the 2-stroke propulsion technology as shown in Table 7.

6.2 EEDI CONSIDERATIONS

The Energy Efficiency Design Index (EEDI) is essentially used to quantify CO_2 emissions from vessels, and defining baselines for new ship designs, which in turn instigates the development of more efficient ship designs. The IMO extended the application of the EEDI to LNGCs in September 2015, with the intent that from September 2015, all new LNGC carriers will need to be 10% more efficient than the approved baseline value, with this percentage rising to 20% by 2020 and finally to 30% by 2025.

A study (Attah & Bucknall, 2015) conducted on the influence of the EEDI on the design of future LNG carriers, showed that the current baseline introduced in 2015 was relatively kind to DFDE vessels when compared to the less efficient steam LNGCs, as one baseline equation was applied to both designs. The study highlighted that an overwhelming majority-97% of current DFDE designs will fall below the 2020 benchmark without recourse to any efficiency improvement technologies, while 56% of current DFDE designs will still fall below the baseline value for 2025. These results are summarized in Figure 7.

Propulsion Type (Nos)	Phase 0 Compliant	Phase 1 Compliant Sept 2015 – Dec 2019	Phase 2 Compliant 2 1 Jan 2020 – 31 Dec 2024	Phase 3 Compliant 1 Jan 2025 – Onwards
Diesel Electric (32)	32- (100%)	32- (100%)	31- (97%)	18- (56%)

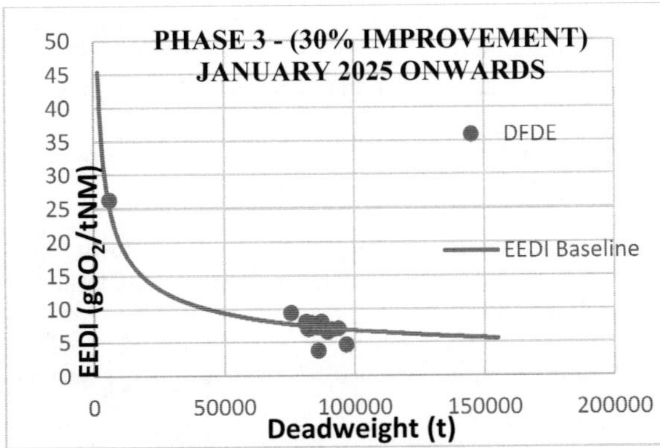

Figure 7: EEDI Analysis of the current DFDE LNGC Fleet. Source: (Olsen, 2016)

Propulsion Type (Nos)	Phase 0 Compliant	Phase 1 Compliant Sept 2015 – Dec 2019	Phase 2 Compliant 2 1 Jan 2020 – 31 Dec 2024	Phase 3 Compliant 1 Jan 2025 – Onwards
Diesel Electric (32)	32- (100%)	32- (100%)	31- (97%)	29- (91%)

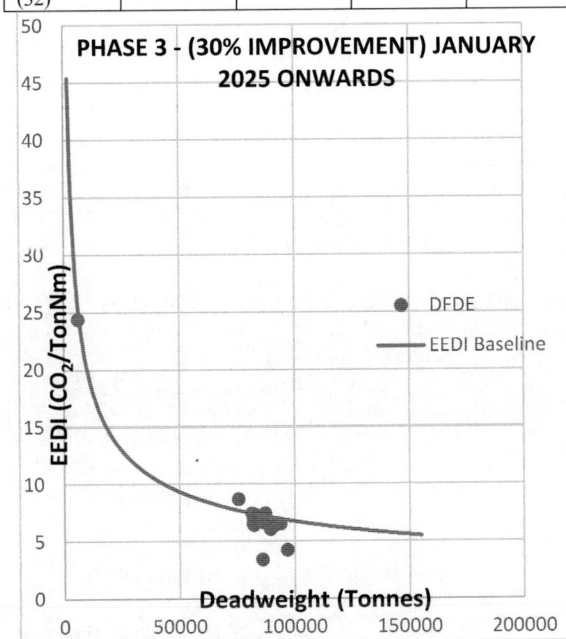

Figure 8: EEDI Analysis of the current LNGC Fleet with Upgraded DFDE systems. Source: (Olsen, 2016)

However, the use of the EGWHRS system further improves the efficiency of the DFDE, by reducing the SFOC component, for when the results of this study were applied to the EEDI analysis carried out in the EEDI study it is seen that this efficiency increase has further increased the number of DFDE vessels falling under the baseline since 91% of designs now fall under the required baseline for 2025 onwards as detailed in the table in Figure 8.

This is particularly noteworthy. Through exploitation of WHRS no other further improvements in technology will be required from phase two as the adoption of the EGWHRS will meet the IMO mandated requirement of the EEDI for new LNGC designs.

6.3 COST IMPLICATIONS OF RESULTS

The second part of the study also captured, the use of the EGWHRS over a month' of operation and compared the savings. From Table 5 it is seen that for vessel 1, the operational profile was such that the TG was in use for 21 out of the 31 days;- essentially 67% of the time. On vessel 2, it was used for 7 days (22%), while for vessels' 3 and 4 the usage was 18 days (58%) and 16 days (52%) respectively.

In order to extrapolate the figures from the EGWHRS trials to the usage on all four vessels, a factor was developed to estimate the gas savings per 1KW of Turbo Generator generated electrical Power.

$$Gas\ savings\ per\ kW\ TG\ Power = {}^{15.3388}/_{1520} = 0.01009\ {}^{m^3}/_{kW}$$

This factor was then applied to the four ships to estimate the fuel savings, over the one month trial period by multiplying the factor obtained in the equation above to the TG readings recorded over the one month period, according to the equations below.

$$0.01009\ {}^{m^3}/_{kW} \times TGPower \times Monthly\ Running\ Hours = Total\ Gas\ Saving$$

Total gas saving was then multiplied by a fuel oil equivalent factor of 0.484 to obtain the equivalent HFO fuel saving to estimate the cost savings based on the cost of Bunkered HFO.

$$Total\ Gas\ Saving \times 0.484 = Total\ HFO\ Equivalent\ Saving (Tonnes)$$

Therefore, based on an average HFO bunker price of \$305/Ton (as at February 2017) (Ship & Bunker Prices, 2017), the total HFO equivalent saving can be estimated according to the formula below

$$Total\ HFO\ Equivalent\ Saving \times 305 = Estimated\ Cost\ Saving\ per\ vessel$$

The full estimated monthly cost summary is represented in Table 8.

Table 8: Estimated monthly cost summary

Vessel 1		Vessel 2	
Total Gas Saving in Month	313.007m^3	Total Gas Saving in Month	94.8m^3
Fuel Oil Equivalent of Gas	151.0tonnes	Fuel Oil Equivalent of Gas	45.8tonnes
Estimated Fuel Cost Savings	**$49,833**	Estimated Fuel Cost Savings	**$15,135**
Vessel 3		**Vessel 4**	
Total Gas Saving in Month	207.7m^3	Total Gas Saving in Month	193.0m^3
Fuel Oil Equivalent of Gas	100.6tonnes	Fuel Oil Equivalent of Gas	93.4tonnes
Estimated Fuel Cost Savings	**$33,180**	Estimated Fuel Cost Savings	**$30,828**

From the above table it can be seen that the use of EGWHRS could realise savings of circa $30,000.00 per month per vessel, with an annual yearly savings of $360,000.00 per vessel. This annual savings when compared to the installation costs of $1.2 million per vessel based on the construction prices in 2015, the savings is expected to recover the equipment costs in 3.5 years. This cost recovery duration is expected to be shorter with higher fuel prices, or if the vessels spend more time at higher service speeds.

7. CONCLUSIONS AND FUTURE WORK

This paper explored the use of EGWHRS as a method to improve the efficiency of DFDE carriers. A case study was presented utilising on-board results. From the results obtained and analysis detailed within this study the following points can be ascertained:

- The use of WHRS delivers an increase in energy efficiency of the DFDE of approximately 6.6% at full load sea going conditions.
- The monthly utilisation analysis carried out on the vessels indicated a savings of between $15,000.00 and $45,000.00 with average savings of $30,000 per month per vessel with an annualized savings of $360,000.00 per year per ship.
- Based on the cost of installation of $1,200,000.00 per vessel the average fuel savings per vessel will recover this additional cost in 3.5 years. The recovery time will inevitably become shorter with higher fuel prices or if the vessel spends a larger amount of time at service speeds.
- At low ship loads the turbo generator becomes unreliable, as there is not sufficient exhaust gas to generate the required minimum steam flow requirements.
- When compared to other LNG propulsion options, a 6.6% increase in the energy efficiency translates to a 48% thermal propulsive efficiency. This puts it on par with 2-stroke LNGC options.
- In terms of the EEDI, this 6.6% energy efficiency increase could ensure that 100% of current designs will satisfy 2025 EEDI targets based on the baseline released by the IMO.

The following future studies are recommended based on the above conclusions:

- Current EGWHRS to be optimised to ensure it is available at lower loads to ensure efficiency improvements can be realised at lower loads. This case study covering different vessels highlighted the fact that these vessels spend majority of their time at lower loads, due to the market dynamics.
- The use of an EGWHRS to generate power should be explored for use with 2-stroke diesel options. This in turn will improve the overall efficiency of the systems and increase the benchmark within which LNG vessels will be measured.
- There is also a need to review the EEDI benchmark for DFDE LNGCs, as it appears that the EEDI for LNGCs can be easily achieved by a vast majority of carriers. Beyond 2025, stiffer regulations would be required to encourage greater innovation. As based on already available technology, 91% of current designs fall under the benchmark to 2025 if EGWHRS is adopted.

8. REFERENCES & NOTES

1. United States Department of Energy (2017) *Liquefied Natural Gas Characterisation.* Available: https://energy.gov/sites/prod/files/2013/04/f0/LNG_primerupd.pdf [10 September 2016]
2. Clarksons Shipping Intelligence Network (2014) LNG Carrier Fleet Available: http://www.clarksons.net/sin2010/register/Default.aspx?sValues=rOpt%3dregister%7crSel%3d3%7cZS0_SHIP_TYPE%3dVG9%7ctitle%3dLNG+Gas+Carrier+Fleet%7c [18 July, 2014]
3. WAYNE, W. S. & HOGSON, M [2006]. *The Options and Evaluation of Propulsion Systems for the Next Generation of LNG Carriers.* Available: http://igu.dgc.dk/html/wgc2006/pdf/paper/add10055.pdf [26 September 2013]
4. Clarksons Shipping Intelligence Network (2014) *LNG Carrier Orderbook.* Available: http://www.clarksons.net/sin2010/register/Default.aspx?sValues=rOpt%3dorderbook%7crSel%3d3%7cZS0_SHIP_TYPE%3dVG9%7ctitle%3d

LNG+Gas+Carrier+Orderbook%7c [18 July, 2014]

5. ADRIANOS, K., (2006) *Techno-Economic Evaluation of Various Energy Systems for LNG Carriers.* Available: http://dspace.lib.ntua.gr/bitstream/123456789/3009/3/andrianosk_ingcarrier.pdf
6. Clarksons Shipping Intelligence Network (2017) LNG Carrier Orderbook Available: https://www.clarksons.net/wfr2/fleet [02 October, 2017]
7. CHANG D, RHEE T, NAM K, & LEE S. (2008) *A study on availability and safety of new propulsion systems for LNG Carriers.* Available: http://ac.els-cdn.com/S0951832008001105/1-s2.0-S0951832008001105-main.pdf?_tid=62906dd2-26f8-11e3-a0be-00000aacb35f&acdnat=1380233579_df20deabaa3077eeeef5702c6cc239c2 [26 July 2014]
8. International Maritime Organisation (2014) *3rd IMO Green House Gas Study 2014* Available: http://allportcargoservices.com/retailnews/allport-knowledge/regulatory-news/international-shipping-cuts-co2-emissions/801684843 [08 August 2014]
9. MAN Diesel and Turbo (2016) *Waste Heat Recovery Systems for Marine Applications.* Available: http://marine.man.eu/docs/librariesprovider6/technical-papers/waste-heat-recovery-system.pdf?sfvrsn=10 [03 October 2017]
10. OLSEN, E. S., (2016) *Optimising Efficiency in DFDE for LNG Carriers.* Available: http://new.abb.com/marine/generations/technical-insight/the-best-of-both-worlds [19 September 2017]
11. SENARY, K., *ET AL* (2016) *Development of a waste heat recovery system onboard LNG carrier to meet IMO regulations.* Alexandria Engineering Journal (2016) 55, 1951-1960. Alexandria August 2016.
12. Samsung Heavy Industries (2015) Machinery Manual of LNG DFDE vessel. Assessed 20th April 2015.
13. ATTAH, E., & BUCKNALL, R., (2015) *Influence and Impact of the EEDI on the Design of Future LNG carriers.* Royal Institute of Naval Architects EEDI Conference. London 22-25 September 2015
14. Ship and Bunker Prices (2017) Worlds Bunker Prices Available: https://shipandbunker.com/prices [30 April 2017]

AUTHORS' BIOGRAPHY

Effiong Ekanem Attah is a final year PhD Student in the Marine Research Group in the Department of Mechanical Engineering at UCL. His current research interests focuses on the technologies currently available to LNG carriers to improve transport efficiencies. Prior to joining UCL, he completed a BEng (Hons) degree in Marine Engineering Technology at the University of Greenwich and an MSc in Maritime Operations and Management at City University London. Professionally, he has worked as a Marine Engineering Officer for Shell/Nigeria LNG onboard LNG carriers with varied responsibilities, and is currently Head of Maritime Projects ashore.

Mfreke Ebong is Technical Superintendent in the Ship Management arm of Nigeria LNG, overseeing the management of DFDE LNG carriers. He has over seventeen years' sea going experience in LNG shipping sailing up to the rank of Chief Engineer. His previous role was as site manager in South Korea overseeing the construction of four DFDE LNG carriers.

Richard Bucknall is a Professor of Marine & Power Systems in the Department of Mechanical Engineering at UCL. His research interests are in the fields of electrical systems used in the marine environment and marine propulsion systems more generally, and he currently supervises eight PhD students and chairs the Marine Research Group. He has extensive sea going experience gained in the merchant navy with BP and working with the Royal Navy, UK MOD and extensive academic experience working at the Royal Naval Engineering College (Plymouth) and at UCL publishing over 130 papers and books. He is also the Programme Director for the MSc in Marine Engineering and the MSc in Power Systems Engineering programmes. Externally, he is the Co-Editor of the Journal for Marine Engineering and Technology.

APPENDICIES – Tables 3, 4 & 5

Table 3: Mode 1 Results

	MODE 1	4Engines Running		
1	SHIP'S NAME:	XXXXXXX		
2	**DATE**	**11-Feb-17**		
4	TEST START	hrs	9:00	
5	TEST FINISH	hrs	11:00	
6	TEST DURATION	hours	2	
7	DISPLACEMENT	tonnes	118504	
8	DRAFT FORWARD	metres	11.26	
9	DRAFT AFT	metres	11.26	
10	SHIP'S HEADING	degrees	270	
11	WIND -	force & direction	3 x SW	
	SEA STATE		3, SLIGHT	
12	SWELL	metres & direction	0.5 x SW	
13	CURRENT	Spd x Dirn	0.6 x N	
14	OUTSIDE ATMOSPHERIC PRESSURE	mbar	1022	
15	OUTSIDE AMBIENT TEMPERATURE	°C	28	
16	S.W.TEMP	°C	29	
17	E.R. TEMP	°C	36	
18	OBSERVED DISTANCE	nautical miles	36.1	
19	SPEED BY OBSERVATION	knots	18.05	
20	SPEED BY LOG (GPS Log)	knots	17.75	
			PORT	STBD
21	REVOLUTION COUNTER AT START		9048359	9024012
22	REVOLUTION COUNTER AT FINISH		9055535	9031189
23	RPM	rpm	72.60	
24	ENGINE DISTANCE	nautical miles	38.40	
25	ENGINE SPEED	knots	19.20	
26	APPARENT SLIP	%	5.99	
			Start	End
27	PORT Shaft Total Power (From Shaft Torque Meter)	kWh	44446.878	44475.059
28	STBD Shaft Total Power (From Shaft Torque Meter)	kWh	41418.754	41446.94
29	No.1 GENERATOR LOAD	kW	6171	
30	No.2 GENERATOR LOAD	kW	6324	
31	No.3 GENERATOR LOAD	kW	6172	
32	No.4 GENERATOR LOAD	kW		
33	No 5 GENERATOR LOAD	kW	6472	
34	T/A LOAD	kW		
35	PORT PROPULSION MOTOR LOAD	kW	11064	
36	STBD PROPULSION MOTOR LOAD	kW	10881	
37	PORT SHAFT POWER	kW	10784	
38	STBD SHAFT POWER	kW	10360	
39	HOTEL LOAD	kW	3194	
40	No.1 MGE GAS METER AT START	Kg	4130437.900	
41	No.1 MGE GAS METER AT FINISH	Kg	4132588.700	
42	No.2 MGE GAS METER AT START	Kg	3625455.900	
43	No.2 MGE GAS METER AT FINISH	Kg	3627631.700	
44	No.3MGE GAS METER AT START	Kg	3429356.900	
45	No.3 MGE GAS METER AT FINISH	Kg	3431492.900	
46	No.4 MGE GAS METER AT START	Kg	1958211.200	
47	No.4 MGE GAS METER AT FINISH	Kg	1958211.200	
48	No.5 MGE GAS METER AT START	Kg	3718990.900	
49	No.5 MGE GAS METER AT FINISH	Kg	3721173.900	

Table 4: Mode 2 Results

	MODE 2	4Engines + T/A Running		
1	SHIP'S NAME:	XXXXXX		
2	**DATE**	**11-Feb-17**		
4	TEST START	hrs	12:00	
5	TEST FINISH	hrs	14:00	
6	TEST DURATION	hours	2	
7	DISPLACEMENT	tonnes	118504	
8	DRAFT FORWARD	metres	11.26	
9	DRAFT AFT	metres	11.27	
10	SHIP'S HEADING	degrees	270	
11	WIND -	force & direction	3 X SW	
12	SEA STATE		3, SLIGHT	
13	SWELL	metres & direction	0.5 X SW	
14	CURRENT	Spd x Dirn	0.6 X N	
15	OUTSIDE ATMOSPHERIC PRESSURE	mbar	1020	
16	OUTSIDE AMBIENT TEMPERATURE	°C	29	
17	S.W.TEMP	°C	30	
18	E.R. TEMP	°C	36	
19	OBSERVED DISTANCE	nautical miles	36	
20	SPEED BY OBSERVATION	knots	18	
21	SPEED BY LOG (GPS Log)	knots	17.65	
			PORT	STBD
22	REVOLUTION COUNTER AT START		9034804	9059153
23	REVOLUTION COUNTER AT FINISH		9066366	9044019
24	RPM	rpm	72.40	
25	ENGINE DISTANCE	nautical miles	38.30	
26	ENGINE SPEED	knots	19.15	
27	APPARENT SLIP	%	6.01	
			Start	End
28	PORT Shaft Total Power (From Shaft Torque Meter)	kWh	44489.016	44517.34
29	STBD Shaft Total Power (From Shaft Torque Meter)	kWh	41460.887	41489.219
30	No.1 GENERATOR LOAD	kW	5879	
31	No.2 GENERATOR LOAD	kW	6119	
32	No.3 GENERATOR LOAD	kW	5758	
33	No.4 GENERATOR LOAD	kW		
34	No 5 GENERATOR LOAD	kW	5897	
35	T/A LOAD	kW	1520	
36	PORT PROPULSION MOTOR LOAD	kW	11066	
37	STBD PROPULSION MOTOR LOAD	kW	10881	
38	PORT SHAFT POWER	kW	10786	
39	STBD SHAFT POWER	kW	10361	
40	HOTEL LOAD	kW	3226	
41	No.1 MGE GAS METER AT START	Kg	4133604.900	
42	No.1 MGE GAS METER AT FINISH	Kg	4135608.100	
43	No.2 MGE GAS METER AT START	Kg	3628644.000	
44	No.2 MGE GAS METER AT FINISH	Kg	3630668.900	
45	No.3MGE GAS METER AT START	Kg	3432522.200	
46	No.3 MGE GAS METER AT FINISH	Kg	3434528.500	
47	No.4 MGE GAS METER AT START	Kg	1958211.200	
48	No.4 MGE GAS METER AT FINISH	Kg	1958211.200	
49	No.5 MGE GAS METER AT START	Kg	3722217.700	
50	No.5 MGE GAS METER AT FINISH	Kg	3724255.500	

Table 5: Vessels EGWHRS-TG Readings

VESSEL 1				
DATE	**TG POWER (KW)**	**RUNNING HOURS**	**DAY'S AVERAGE SPEED**	**COMMENTS**
01/05/2017	1630	24	17.92	Laden
02/05/2017	1554	24	17.75	Laden
03/05/2017	1460	24	17.5	Laden
04/05/2017	1428	23	17.65	Laden
05/05/2017	1521	24	17.63	Laden
06/05/2017	1062	24	16.92	Laden
07/05/2017	0	0	0	Transit Suez
08/05/2017	0	0	0	Transit Suez
09/05/2017	0	0	7.72	Slow steaming
10/05/2017	0	0	0	Slow steaming
11/05/2017	0	0	0	Slow steaming
12/05/2017	0	0	0	Slow steaming
13/05/2017	0	0	0	Slow steaming
14/05/2017	0	0	0	Slow steaming
15/05/2017	0	0	0	Discharging
16/05/2017	0	0	0	Discharging
17/05/2017	1565	0.1	16.33	Ballast
18/05/2017	1497	24	20	Ballast
19/05/2017	1632	24	19.79	Ballast
20/05/2017	1610	24	19.75	Ballast
21/05/2017	1596	24	20.46	Ballast
22/05/2017	1589	24	17.46	Ballast
23/05/2017	1450	24	17.5	Ballast
24/05/2017	1635	24	19.42	Ballast
25/05/2017	1636	24	20.04	Ballast
26/05/2017	1636	24	20.21	Ballast
27/05/2017	1607	24	19.29	Ballast
28/05/2017	1557	24	17.75	Ballast
29/05/2017	1620	25	18.08	Ballast
30/05/2017	1567	24	19.54	Ballast
31/05/2017	1518	24	19.8	Ballast
VESSEL 2				
DATE	**TG POWER (KW)**	**RUNNING HOURS**	**DAY'S AVERAGE SPEED**	**COMMENTS**
01/05/2017	0	0	11.6	Ballast
02/05/2017	0	0	11.8	Ballast
03/05/2017	0	0	10	Ballast
04/05/2017	0	0	9.8	Ballast
05/05/2017	0	0	10.1	Ballast

06/05/2017	1593	16	17	Ballast
07/05/2017	1472	24	15.4	Ballast
08/05/2017	1537	14	16.6	Ballast
09/05/2017	1499	24	17.6	Ballast
10/05/2017	1409	25	19	Ballast
11/05/2017	1461	24	18.9	Ballast
12/05/2017	1502	24	18.5	Ballast
13/05/2017	0	0	0	Vessel in port
14/05/2017	0	0	0	Vessel in port
15/05/2017	0	0	16.7	Laden
16/05/2017	0	0	14.46	Laden
17/05/2017	0	0	14.38	Laden
18/05/2017	0	0	12.04	Laden
19/05/2017	0	0	12.08	Laden
20/05/2017	0	0	14.54	Laden
21/05/2017	0	0	12.13	Laden
22/05/2017	0	0	11.5	Laden
23/05/2017	0	0	12.46	Laden
24/05/2017	0	0	12.13	Laden
25/05/2017	0	0	12.91	Laden
26/05/2017	0	0	11.71	Laden
27/05/2017	0	0	12.08	GCU is on
28/05/2017	0	0	13	GCU is on
29/05/2017	0	0	12.5	GCU is on
30/05/2017	0	0	12.88	GCU is on
31/05/2017	0	0	12	GCU is on
VESSEL 3				
DATE	**TG POWER (KW)**	**RUNNING HOURS**	**DAY'S AVERAGE SPEED**	**COMMENTS**
01/05/2017	1147	23	15.875	Ballast
02/05/2017	1499	25	17.92	Ballast
03/05/2017	1477	24	17.5	Ballast
04/05/2017	1491	25	17.36	Ballast
05/05/2017	1072	24	17.04	Ballast
06/05/2017	931	25	15.48	Ballast
07/05/2017	948	24	15.42	Ballast
08/05/2017	897	25	15.64	Ballast
09/05/2017	1287	24	15.88	Ballast
10/05/2017	1492	24	17.96	Ballast
11/05/2017	1388	25	17.32	Ballast
12/05/2017	1493	24	16.96	Ballast
13/05/2017	1102	24	17.6	Ballast
14/05/2017	1116	24	17.83	Ballast
15/05/2017	969	24	16.16	Ballast

16/05/2017	981	24	16.32	Ballast
17/05/2017	936	24	16.16	Ballast
18/05/2017	0	0	15	Load port
19/05/2017	0	0	0	Load port
20/05/2017	0	0	15.83	Slow steaming
21/05/2017	997	2	10.6	Slow steaming
22/05/2017	0	0	9.42	Slow steaming
23/05/2017	0	0	9.5	Slow steaming
24/05/2017	0	0	9.2	Slow steaming
25/05/2017	0	0	12.5	Slow steaming
26/05/2017	0	0	9.03	Nil
27/05/2017	0	0	8.6	Slow steaming
28/05/2017	0	0	8.5	GCU is on
29/05/2017	0	0	8	Slow steaming
30/05/2017	0	0	8.17	Slow steaming
31/05/2017	0	0	10.17	Slow steaming
VESSEL 4				
DATE	**TG POWER (KW)**	**RUNNING HOURS**	**DAY'S AVERAGE SPEED**	**COMMENTS**
01/05/2017	1373	24	17.46	Laden
02/05/2017	1328	24	17.17	Laden
03/05/2017	1261	24	17.25	Laden
04/05/2017	1205	24	16.5	Laden
05/05/2017	1238	24	16.37	Laden
06/05/2017	1229	24	15.67	Laden
07/05/2017	1148	24	14.71	Laden
08/05/2017	1138	24	14.48	Laden
09/05/2017	0	0	0	Discharging
10/05/2017	1306	6	17.78	Ballast
11/05/2017	1320	24	17.28	Ballast
12/05/2017	1380	24	18.3	Ballast
13/05/2017	1380	24	17.37	Ballast
14/05/2017	0	0	13.87	Slow steaming
15/05/2017	0	0	9.17	Slow steaming
16/05/2017	0	0	8.62	Slow steaming
17/05/2017	0	0	8.87	Slow steaming
18/05/2017	0	0	8.62	TG is off
19/05/2017	0	0	7.08	TG is off
20/05/2017	0	0	6.54	NIL
21/05/2017	0	0	6.87	TG is off
22/05/2017	0	0	7.21	NIL
23/05/2017	0	0	7.75	TG is off
24/05/2017	0	0	7.67	NIL
25/05/2017	0	0	7.25	TG is off

26/05/2017	0	0	12.67	NIL
27/05/2017	0	0	0	Loading
28/05/2017	1280	14	18.64	Laden
29/05/2017	1330	24	18.5	Laden
30/05/2017	1344	24	18.21	Laden
31/05/2017	1318	24	18.08	Laden

DRIVING RELIABILITY, EFFICIENCY, AND SAFETY ON LNG/LPG VESSELS THROUGH AUTOMATION

B O Berntsen and **K Høglund,** Høglund Marine Automation AS, Norway

SUMMARY

As the popularity of LNG as a marine fuel grows, the LNG carrier segment continues to expand, while a new generation of vessels is emerging to meet the growing demand for LNG bunkering services. Høglund is pioneering the automation of these vessels, and has fitted 15 LNG-powered ships with its automation solutions, which represent 12% of the global LNG-powered fleet. While automation system interfaces and design have evolved to fit big scale LNG carriers over the last four decades, medium and small-scale LNG carriers/bunker vessels is a new market, which is still not optimized. This opens up the opportunity for new ideas and fresh thinking to ensure success in the automation of these new vessel types. But as always with new technologies, reliability, efficiency, and safety must be prioritized to make sure systems are working as intended, while still reducing capex and opex.

NOMENCLATURE

CTS	Custody Transfer System
DP	Dynamic Positioning
FDS	Functional Design Specification
FO	Fuel Oil
GCU	Gas Combustion Unit
GVU	Gas Valve Unit
HAT	Harbour Acceptance Test
HMA	Høglund Marine Automation AS
HMI	Human Machine Interface
HW	Hardware
I/O	Input/ Output signal
ICAS	Integrated Control and Alarm System
LNG	Liquid Natural Gas
LPG	Liquid Petroleum Gas
LTS	Long Time Support
MG	Main Generator
MSB	Main Switch Board
MTBF	Mean Time Between Failure
MTTR	Mean Time to Repair
OPC	OLE for Process Control
OS	Operator Station
PLC	Programmable Logic Controller
PMS	Power Management System
PSV	Platform Supply Vessel
SAT	Sea Acceptance Test
SW	Software
TCO	Total Cost of Ownership

1. INTRODUCTION

Modern Integrated Alarm and Control System (ICAS) on board LNG/LPG vessels consist of thousands of input and output signals with widespread signal interfaces and functions. The ICAS is wired together to solve many tasks, which alone can be trivial, but put together become complex. HMA has always focused on splitting up complicated tasks and implementing them as series of easy tasks - an approach, which has a track record of success in all our projects. We structure the complete project in lists where all similar functions are organized together. This gives a good overview of the complete process and breaks down the tasks, which enable more powerful development.

1.1 NEW BUSINESS

When new business segments are developed, it's important that all participants are responsive to what can be inherited from other segments and which components must be replaced or reengineered to fit the new segment. Høglund has extensive experience from several fields within marine automation; oil rigs, passenger ships, chemical tanker, ferries as well as advanced offshore construction and seismic vessels. This background makes Høglund a competitive player in the new small-scale LNG segment, which is rapidly growing at present time. This paper will discuss methods of designing marine automation systems with response to overall philosophy, segmentation, equipment under control, signal flow, logging and data distribution while ensuring total cost of ownership is kept low.

1.2 PRIMARY DESIGN CONSIDERATION

When designing a commercial vessel, automation is often not considered in detail by the design company responsible for the basic ship design. However, the technical specifications ship owners use against ship yards to get a tender on a vessel are often partly written by the design company. Because time is a critical component, many of the more general aspects of a ship are just copied from previous specifications. Examples of this practice can be cabin facilities for crew, or even the automation system. The result is that the shipyards, which are also competing against each other, will quote a ship per specification, with the cheapest possible automation system.

1.3 ICAS DESIGN PHILOSOPHIES

In the current market, there are two prevailing design philosophies. These philosophies are both focusing on getting the lowest possible cost of a complete ICAS system while maintaining reliability, safety, and efficiency of the vessel. The main difference is the means each philosophy uses to achieve its goals. The industry is under constant cost pressure to deliver cheaper and more reliable ships.

1.3 (a) Segmented Approach

To avoid headaches during commissioning, many of the systems are delivered as "turn key" systems where a pump manufacturer or a valve manufacturer delivers their own proprietary control system. A serial line will then provide the link to the Integrated Alarm and Control System. This way of designing a system has its advantages – such as segmented commissioning responsibilities between the ICAS and valve supplier, which can run in parallel, leaving only the interface between the two to be tested before the whole plant is complete. Splitting up the various systems allows for competition between the subcontractors of the yard. The shipyard can easily manage the risk since changing out a system even during construction will be possible, although undesirable.

1.3 (b) Integrated Approach

Another approach will be for the shipyard to organize all automation functions and specify them for the chosen ICAS supplier to implement them, so that the ICAS supplier is only delivering I/O modules, CPUs, and HMI. Sensors and actuators can then be purchased from other vendors, which specialize on their domains and therefore are delivering equipment both cheaper and with higher quality than suppliers that try to focus on everything. Signal interfaces between sensors and ICAS need to be standard for this approach to work and this makes competition between ICAS suppliers more intense since there are many competitors utilizing standard industrial HW.

Control functions are realized directly in the ICAS, which makes for easier debugging and modifications if parts of the system under-perform during HAT or SAT tests. Less change management cost makes systems customized to their use cheaper.

The disadvantage with this approach is that the shipyard needs to develop detailed FDS of the different control functions, which is more labour intensive than buying a standard control package for a specific system.

We have now given a very brief explanation of ICAS design philosophies. Both are well applied within the marine industry. We will now go into more detail to discuss design philosophy choices when applying automation to new business segments, like small-scale LNG carriers or ships using LNG as fuel.

2. TODAY'S SITUATION

2.1 INCREASED COMPUTER UTILIZATION

Many consider the shipping industry to be conservative. This also applies to automation. Concepts widely applied throughout the industry have been inherited from the days when hardware logic and relays were controlling the plant. The industry sticks to these ideas. The reason is that they are easy and cheap to repair and on board competency to troubleshoot and to solve issues are readily available.

When a ship is sailing, the available troubleshooting and repair capabilities are what you have on board. The crew must therefore have the necessary training and tools together with spare parts to be able to fix problems hands on.

As PLCs became cheaper than conventional relays, and cost of engineering and commissioning was reduced by increased utilization of PLCs, more suppliers are installing a PLC in their systems. But it has its challenges:

Ship owners report the following issues:

- Huge increase in unexplained events from various sub-systems, sometimes with serious damage.
- Undocumented PLCs are "found" inside faulty systems for which service is hard to get.
- Faulty PLCs can put modern vessels out of operation for longer periods of time.
- The number of "sub control systems" are increasing and creating huge challenges.
- Difficult or impossible to recreate problems and find bugs.
- The crew is searching blindly in the system for the cause of an event, and desperately looking for some reason and actions to report.
- Systems need a long and troubling commissioning period, before they are performing to a minimum standard.
- The crew now often accepts poor systems if they can get around the problem.
- The occurrence of critical side effects after a SW upgrade has become common; therefore systems with poor performance and known bugs are accepted if work-arounds are possible.
- Nobody has the full knowledge of a modern computer system.
- System upgrade has become a challenge, and sometimes a nightmare.
- Replacement of computer parts from stock has become almost impossible.

It's clear from these reports that serious issues exist. Why are the on board crew unable to correctly troubleshoot and repair the plant? Are they unwilling? Or is their training insufficient?

2.2 DOCUMENTATION

2.2 (a) Hardwired Connections

All wires and cables connected on board a ship are normally well documented. Cables are numbered according to which system they are a part of and interface connections are documented either in a list or by a drawing. This documentation facilitates a way to troubleshoot a plant in an easy and convenient way using a multi-meter as the main tool. Most faults can then be addressed and corrected by using the tools available to any electrician. There is a consensus about the need to have updated drawings showing actual connections.

2.2 (b) Software Connection and Logic

While there is a broad consensus about the need to have documentation and diagrams in HW systems, no real effort has been put to get access to actual program running in PLC's on board ships. Ideally, the information technicians had before though the wiring diagrams should have been available on board either as printouts or online troubleshooting capabilities of the running systems. Providing information gives the crew a real possibility of finding out what's happening and to figure out the best remedy. So why is the information that crew on board need not available to them? The answer is in many cases unclear, but here are a few reasons why.

One potential reason may be that it is necessary for a particular cable, laptop, or a piece of software to be used to access the systems in question and in some cases, there is no way to protect the system against changes if access is granted. This makes the suppliers reluctant to facilitate any online mechanism for on-site troubleshooting since incorrect changing of parameters can cause serious damage to the plant and costly warranty claims.

There are also suppliers that have a service philosophy, which aims to get a certain per cent of their total revenues from service jobs and since ship-owners need to pay for service after the warranty period, on board service becomes a valuable revenue stream.

It is also a question of protecting the intellectual properties of the software making it hard to copy and read since software piracy is becoming a major concern in shipboard automation.

The advantages of installing a small PLC in a local valve controller, for instance will be eaten away if the functions of such a system will be as good as an ICAS system where all the troubleshooting tools normally is embedded in the system. As proper online debugging tools are costly to develop and deliver with the system, these are often omitted.

2.3 SEGMENTED APPROACH

2.3 (a) Definition

We define a segmented architecture, to be several types of PLCs controlling their own equipment under control in the plant, and where signals are sent to the ICAS via a communication line. SW can be standardized by the component supplier, but not always.

2.3 (b) Benefits

There are obvious benefits from letting each component supplier control his own component.

First, it is the definitively the cheapest method with respect to purchase and installation, as this control system is already developed and many deliveries have been done. This means that it most likely has proved to work, and it should not raise any problems during commissioning.

Often the control system comes ready mounted and wired to the components, and this saves the yard for many concerns:

- No I/O list to be made by the yard.
- No I/O checks to be performed.
- Function test may be done before delivery, and the yard will save time and efforts.
- The component supplier will develop tailored SW based on detailed knowledge of the product.

It is also possible to claim that failures of the actual system are easier to determine, as the interface is clear due to defined signal transmissions to the superior ICAS system.

2.3 (c) Disadvantages

As the Software tradition within shipping is still "young", there is no established way for the owner or yard to verify that the supplied component, with its "standard" control system is working, as it should in all situations.

As a component supplier is often relatively inexperienced in SW development, there are a lot of bugs in the SW, while revision history and change management procedures are often lacking. The statistics show that a considerable part of the problems during commissioning are related to SW problems. That is not to say that this only occurs in the small independent PLC's delivered by subsystem vendors - but that superior ICAS systems are often more open to debug and tools for programming are available from the HMI.

2.3 (d) SW Modification and Debugging

As the SW is stored in a proprietary PLC, and the source code for the unit is not available, it is not possible for any other representative but the component supplier to do SW modifications or parameter adjustments.

This is often a major problem, as it will require the engineer for the component supplier to be present to correct small parameter adjustments or SW adjustments due to interface differences from the standard delivery.

If a serious problem occurs during operation, it is normally not possible for the on board electrician or engineer to go into the system and make a temporary modification to proceed with operation.

It might be that an interlock signal is failing, but the actual blocking is confirmed as OK, and the crew need to proceed with the operation, but a blocking cannot be done from the superior ICAS system, and only by a service engineer that will need days to attend the vessel.

If this had been done in the ICAS, it probably could have been done by the crew with a standard function, or via a remote connection by the ICAS supplier.

2.3 (e) SW Verification and version control

Unfortunately, many smaller suppliers of PLC programs do not have the necessary skills and experience with advanced computer systems.

Most people think that a thorough SW test will ensure safe operation, but that is unfortunately not true.
A SW test does not reveal all the possible bugs in the SW, it only demonstrates that it did work during the actual test. As there are almost infinite combinations in the SW it is impossible to trust a simple test.

And unfortunately, there are no formal requirements for a PLC SW programmer, like there are for a welder or an electrician.

In addition to this, the SW programmer's job cannot be overseen effectively, as nobody today is inspecting the SW.

Another problem is the version control. Often there is no information on this available on board the vessel, and any new versions may have been loaded after sea trial. The new SW should have been tested and checked as to previously entered parameters, and possible new functions that were implemented during harbour and sea trials, - however, this is not always the case.

2.4 INTEGRATED APPROACH

2.4 (a) Definition

Generally, the word "full integration" can be used, when: The same system controls the actual part of process, with the same components, the same SW modules and controlled from the same HMI, with the same operational functions as the overall ICAS.

There are some questions that can be raised to find the degree of integration of a system:

- Does the system communicate with the same communication protocol?
- Are the components of the same brand, and model? I.E. uses the same spare parts?
- Does it have the same SW architecture?
- Does it use the same SW tools for programming?
- Does it use a common SW library for the programming?
- Are all properties in the database available from the ICAS HMI?
- Is it possible to set ALL parameters in the system from the ICAS HMI?
- Is it possible to bring up the different SW blocks in the system from the HMI for debugging?

Even if not all this is true, the system may have a certain degree of integration, but it is not done to a full extent.

2.4 (b) Benefits

There are many benefits to selecting an integrated system:

- One single company is responsible.
- Single point of contact.
- Same HW.
- Same HMI.
- Same spare parts.
- Same training courses.
- Detailed I/O information in screen (as all is in the main system) .
- Easy logging (as all data is in the main system).
- All parameters are stored in a central database.
- One single person can debug all the way down to I/O level.

2.4 (b) Standard Hardware

Historically, it was advantageous to develop specialized hardware to control the processes on board a ship. The reason for this was, in part, that there were special requirements for signal interfaces against process and that the processes to be controlled on board a ship were of a simpler nature than that which was to be controlled in a factory shore-side. At the same time, the requirements for operational safety of a ship have been very strict as one cannot stop the process and go home if something stops working. A ship is like an island, where everything from power supply to food production and propulsion systems must be in operation to ensure ship, cargo and crew are safe. In addition, the ship must do the task for which customers have paid.

As the maritime industry is constantly hunting to find cheaper ways to acquire ships that meet both regulatory requirements as well as customer and classification societies, it opened an opportunity for companies that created proprietary control systems, specially designed for marine use.

As industrial suppliers of PLC systems noticed the potential to expand into the marine market, they went through the process of approving their products for use on board ships. This development revealed the fact that systems used shore side could be installed on ships using the same hardware and software platform, which in turn increased the quality and reduced the cost of the automation system. The only downside was that flexibility went down as one had to deal with the signal interfaces used in the industry, but this downside was very well compensated as shore-based industry control system requirements are very strict and reliability is paramount to a well-functioning factory.

2.4 (c) Standard Software

To reduce the likelihood of errors and to further develop software, it is a great advantage to use standard software to control standard processes. Dividing up software packages induces the need for interfaces between software components, which can be painful, but defining clear interfaces is also a factor which in practice gives better control of the software and increases portability.

2.4 (d) Challenges with Integrated Systems

An increasingly number of components come with a control system made by the component supplier. As each supplier selects his own system platform, it is not that easy for a yard/owner to get the same system all over the vessel.

To achieve commonality, each component supplier must remove the standard control system before delivery. Then the selected main system integrator must implement new control algorithms based upon the component supplier's specification. The component supplier often rejects doing this, based upon the fact that he does not trust the system integrator the task of protecting his components, and the customer will have to accept limited warranty on the package.

The result of this is that the cost of the integration rises to an impossible level, the same time as the benefits from doing this integration is very unclear for most owners.

Consequently, there is an increased segmentation of system controls, as more component suppliers are supplying their own control system. The "Integrated System" is made up of numerous communication lines to the main ICAS.

Another problem is the fact that no system integrator can hold the detailed knowledge of every component or processes on the vessel. This is only possible for the manufacturer of the actual component. If this actual component is to be Integrated, it will require extensive cooperation between the component supplier and the system integrator.

But this again raises another ever-bigger problem:
The SW contains information that must be protected by the component supplier to protect intellectual property rights. It is not possible for a system integrator to be enlightened of certain smart system functions, without risking revealing this knowledge when implementing the similar control functions for a competitor in a different project. This is today the most obvious reason component suppliers refuse to give the control information to a system integrator.

3. ICAS DESIGN PHILOSOPHY

3.1 TIME UTILIZATION

When developing strategies on how to give the customer the best possible product, it is important to consider the utilization of your employees. With simple analysis, it will be apparent that to achieve the best possible utilization of time it is necessary that all bottlenecks of production are located and addressed.

An example of such a bottleneck may be that when installing an automation system aboard a ship, the shipyard will create a I/O-list of signals going in and out of the automation system and what functions they should have. Once this document has been created, it will be forwarded to the automation provider, whom has the task of setting up systems that can handle input signals and give the correct outputs based on the functions specified in the list. If this list is received and used directly in generation of program code, bottlenecks are being removed, compared to the task of manually synchronizing the list from the shipyard with the functions of the automation system. Another advantage is that ICAS lists can be completed on a later stage in the project, making changes less likely. Change management can become a major cost when working on pilot projects like small-scale LNG. But there are many reasons why most automation suppliers do not generate the program code based on the shipyard I/O-list. One is the fact that most shipyards uses different format on their list and functions are specified differently, making decoding an issue. A standard format in use as an interface between shipyard and automation supplier reduces likelihood of misinterpreting and reduces total workload on the project, reducing cost and increasing quality as an effect.

An ICAS supplier utilizing import tools that can take the I/O-list directly from the shipyard and import it into the ICAS, thus eliminating this bottleneck will have a significant advantage compared to a supplier depending on manually synchronizing the two lists. Another advantage of this approach is that modifications to the I/O-list will be easy to update with the correct details per the requirements of authorities and ship-owners that will arise during the construction period.

The commissioning can also be divided into multiple steps where cable and termination lists can be made earlier in the project and I/O-checks can be performed concurrent with connection of physical signals. If the automation supplier also facilitates the possibility of the ship yard personnel to do I/O-checks, the automation supplier may reduce their effort on site. ICAS service engineers may also set up a remote connection to the system, enabling them to work from the comfort of their offices. Travel and commissioning cost for the plant is reduced, while ICAS engineers are utilized in a more efficient way which in turn leads to reduced cost for both shipyard and owner.

3.2 FAILURE MODE AND AVAILABILITY

To ensure that a PLC-based control system work and fail as intended, there are several challenges that must be addressed in conjunction with design and commissioning of a system. As we all know, hardware units will eventually fail, and the task then will be to limit the effects of such a fault. By analysing the effect of different HW faults, it will be clear which dependencies the different parts of the plant have between each other. During system design it is crucial that these issues are addressed to give the intended result, generally, as few as possible dependencies should be allowed.

3.2 (a) Failure Mode

Another aspect to consider is failure mode. In shore-based industries where most PLCs originate, fail state will often be defined, to which the system will return to if something goes wrong. Most often, this state is stop, but on board a ship it can be dangerous to go to stop if, for instance, the vessel is in a critical part of the operation such as manoeuvring to the dock or navigation in narrow waters. It is important to define to which conditions the system should go to if something goes wrong. In many cases like a PMS, the safest way is to do nothing, a system fail will generate an alarm, but the system configuration will remain unchanged.

3.2 (b) Availability

As we know from availability calculations the availability of a complete system in series is the product of all availability numbers, the product will always be lower than the individual components.

Let's consider the following example with a DP system of an LNG bunkering vessel operating in enclosed harbours. The DP system is connected to PMS, which in turn is connected to MSB.

Given:
$A_{PMS} = 99\%$
$A_{MSB} = 99.98\%$
$A_{DP} = 99.99\%$

$A_{tot} = A_{DP} * A_{PMS} * A_{MSB} = 98.89\%$ which corresponds to 3.8 days' downtime in a year.

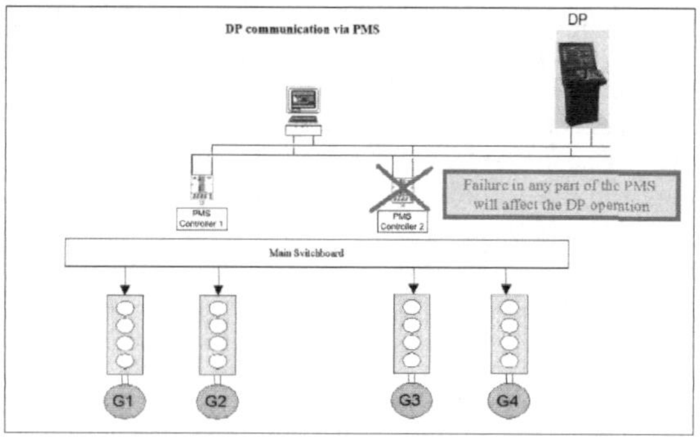

If we instead connect DP system directly to MSB we will get these numbers:

$A_{tot} = A_{DP} * A_{MSB} = 99.997\%$ which corresponds to 2.6 hours' downtime in a year.

A simple design change made a big impact on the result. This example underlines the importance of giving a thought to signal flow and dependencies and how it affects the availability of the total plant.

3.3 SPARE PART AVAILABILITY

As marine automation systems operate on ships sailing worldwide, it is important to ensure that access to spare parts is sufficient. This is best achieved either by a supplier agreement, ensuring delivery time and availability in the ports the ship operates in, or that components that are available worldwide because they are also in use in shore-based industries are being used. The last option only requires one to make an active choice when selecting a vendor, while the first option requires continuous monitoring of system vendor service agreements. Another consideration that must be made is the design life of the plant. Ideally it would be cheapest to buy a plant, which can be maintained for the complete lifecycle of the vessel. This reduces the likelihood of a costly retrofit of the whole ICAS at any point. Since many modern electronic components have a shorter design life than a ship it is worth considering making a lifecycle management assessment of the automation plant.

Generally, PLC's from well-reputed international vendors like Siemens, ABB and Honeywell have a LTS philosophy, which ensures systems can be maintained at least 20 years. When it comes to HMI computers these have a generally lower support life and spare part availability is generally poor beyond 10 years. Selecting a vendor with a good track record in lifecycle management of the plant will be beneficial. To ensure possibility for retrofit at a later stage, signal interfaces and sensor types should be selected according to industry standards to reduce costs if retrofit is necessary during the lifetime of the vessel. Using standard communication methods like OPC between PLC's and OS increases possibilities for part retrofits as apart of mid-lifecycle upgrade of vessel.

4. ICAS VENDOR SELECTION

4.1 GENERAL

To select the correct vendor out of the many who are available, a set of tools come in handy. Assessment of TCO can be performed as well as assessing the owner benefits of choosing a particular system vendor. The availability for the system to feed data to other systems such as maintenance systems, loading computers and owner specific databases containing big data from all ships should also to be considered

4.2 TOTAL COST OF OWNERSHIP

There are several factors that affect the total cost of ownership of an automation system. The most obvious factors are purchase price and annual maintenance cost.

While initial purchase cost is easy to compare, annual service costs are more difficult to compare as different vendors promise a certain service level without necessarily committing to it in a contract. However, there are some things to look out for to make sure service costs are kept as low as possible. By using standardized hardware components available from well-reputed industrial equipment vendors, cost is kept low and quality and availability is kept high. The effect of using standard off-the-shelf hardware induces a reduced flexibility in the use of special signal interfaces, but the advantages are clear. On the software side there are also clear positive effects of choosing a standard platform instead of custom made software structure tailored to ship applications. The things to look for to get as low as possible annual service costs are as follows:

- A reduced spare parts inventory is an obvious effect of standardizing an automation plant. Less spare parts needed on board reduces necessary storage space as well as it increases the likelihood of crew finding the correct spares when needed. It also ensures they are familiar with the equipment in such a way that they can troubleshoot and replace faulty units.
- Using standard software components based on IEC61131 code language also utilizes the maintenance staff in a better way, making them able to troubleshoot more effectively.
- Reduced on-boarding time for maintenance staff coming from other shipping companies and other automation segments is also a clear advantage of using standards. The reduced on-boarding time reduces training cost and makes sure that shipping companies can source crew with the right competence at the right time.
- When using standards, it is also easier to get third party support if there is an issue with the system. The use of third party service staff makes sure service cost rates is kept within industry standards.
- By using standard data transmission protocols interconnection to maintenance/operational and financial systems are commissioned cheaper and more with higher reliability.
- The availability of remote support reduces MTTR as well as reducing travel cost.

Consider the following example of an automation system delivered from Vendor A, B, and C:

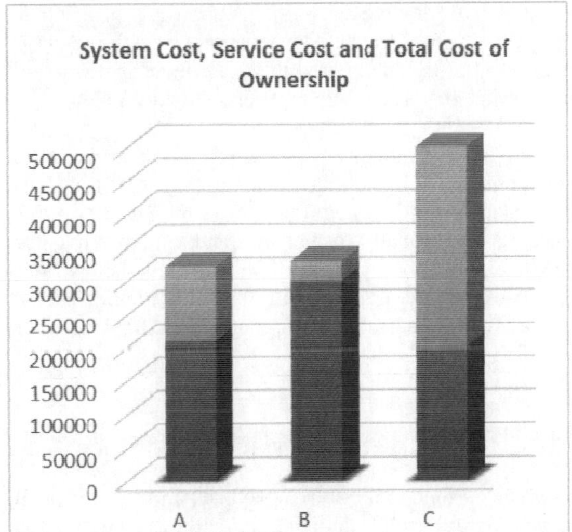

Vendor	System cost [USD]	Service cost [USD]	TCO [USD]	Service cost [USD/Y]	Design Life [Y]	TCO/ Year
A	210000	112500	322500	7500	15	21500
B	300000	30000	330000	2000	15	22000
C	200000	300000	500000	20000	15	33333

Looking at the charts it is obvious that selecting vendor A gives us the lowest TCO, while vendor B gives us the lowest service costs and vendor C has the lowest outright system purchase price. Using TCO as the only criteria, vendor A will be selected, but if initial financing is an issue, vendor C can be selected. The best choice here would be to select vendor B, since service costs are the lowest. The reason is that service requests often are triggered by system failure which, when selecting a vendor with high service cost induces an increased MTTR and a decreased MTBF which reduces the total availability of the system and the operational availability of the vessel is being reduced. This is often impacting the total revenues of the shipping company in much higher proportions than automation system TCO.

4.3 DATA LOGGING

As a superior control system has access to all properties in the database of any signal in the system, it is often an easy task to set up a detailed log of the I/O signals.
A sub-control system may today connect to hundreds of sensors, and control many important outputs. Each of the I/O signals contain a lot of properties, such as Time Delay, Alarm Limit, Filter Factor, Hysteresis etc.

To save money on the interface to the ICAS, only a few of these signals are selected for the communication, and thus available for presentation. Logging a debugging in ICAS, raw values from IO are often not transferred. This is often limiting the possibilities to investigate actual actions and possible SW bugs after an incident.

Data logging of signals has increased both in terms of the time domain and by adding a wider spectacle of variables to the data logging system the later years. Earlier on this data was used mainly to display trend windows and for playback functionality, a function which lets the operator play back historic events using the same mimics as the operator station normally uses, much like a PVR decoder on a TV.

4.4 PERFORMANCE OPTIMIZATION

As much as functionality like playback and trending is a very powerful troubleshooting tool, the data stored is now also used to optimize energy consumption and to propose optimum operation of vessel power plant.

Another aspect of this development is that ship operators gather data from both their ship operations as well as logistics and freight departments, and use this data to optimize their revenues.

5. SPECIFICATION GUIDELINES

5.1 GENERAL

To get the best possible automation plant on board a vessel there are a few guidelines that can be followed when specifying automation plants to ensure that necessary thought is behind the selection. After delivery of the vessel, the Yard should provide a list of all systems containing a PLC.

5.2 HARDWARE REQUIREMENTS

As we will have to live with different types of control systems in a plant, we will need to set basic system requirements corresponding to the level of integration we desire.

A PLC will eventually fail and if, after 10-20 years the suppliers are not able to provide spare parts, the whole system will have to be replaced. Depending on whether the previous vendor used standard software, it might be possible to port the software to run on a different PLC. This is however dependant on available backup of the actual system, and this backup must of course be extracted from PLC before it's failing and it is also only possible using PLC manufacturers programming cables and software.

Since a PLC is cheaper than relays, all suppliers will generally want to control their components with a PLC, but this must be limited to a certain degree. The specification should define where a PLC is accepted and not.

A common PLC brand should be specified if possible. The owner should try to keep the PLC brands as standard as possible to avoid difficulties and high costs in future maintenance.

5.3 SUB-SYSTEM PLC CRITERIA

During development of the specifications as well as under ship construction, the owners need to execute control over which control systems that are allowed containing PLC´s and which are finally chosen. Typical questions should be:

- Why a sub-system needs a PLC based control system on its own?
- Why can it not be controlled from ICAS with standard I/O?

- How will the system be supported during service?
- Is there an obsolescence plan?

Answers justifying the installation of a sub-system PLC:

- The number of signals connected > 100
- The unit contains advanced logics made by the equipment supplier
- The unit has special demands for safe operation, and shutdown
- The system requires local operation independent of the IAS (redundancy aspect).
- The process logic is extensive and is developed under another brand.
- The commissioning engineer from the supplier must be able to make SW modifications.

5.4 SOFTWARE REQUIREMENTS

When a subsystem is accepted, there should be set some basic requirements for all SW. Following items to be delivered along with the system:

- Backup and Listing of the installed SW to be provided
- SW maintenance routines to be available
- SW Version Number
- List of ALL parameters after delivery (if any)
- A complete I/O list of all signals with detailed properties to be provided
- SW code listing (file, PDF or source code)
- FDS of the functions in the delivered system

5.5 MAINTENANCE REQUIREMENTS

- All HW must be available off the shelf for (10-15 years)
- List of all installed HW with supplier information
- Spare batteries (if any)
- Special tools for parameter or SW Adjustment / Upload

5.6 INTERFACE REQUIREMENTS

- No common alarm should be accepted from a standalone system with multiple alarms.
- If a common alarm from sub system containing multiple alarms must be accepted, the output should at least pulse every time a new alarm is activated to notify the ICAS about the new alarm.
- Standard communication bus to be available. (If applicable)
- A list of all available signals for superior communication with defined addresses. (If applicable)

6. CASE STUDY MV CARDISSA

6.1 GENERAL

MV Cardissa is a LNG bunkering vessel operating for Shell Western LNG. This is a new type small-scale LNG vessel, which has a cargo capacity of 6800 m³ of LNG. The total length is 120m and the beam 19 m. The shipyard building the vessel was STX, Korea and ship was delivered spring 2017.

6.2 PROJECT EXECUTION

Høglund Marine Automation delivered the ICAS system together with PMS for this vessel and since it was a new development, it was necessary to change many details from initial specifications.

While most of the components installed on board was standard components, the way they were to be used differentiated from standard operation principles. HMA used its software knowledge to customize the system to give the operators the overview of the plant while ensuring control loops were working as intended.

One example of this method of working is the way fuel consumption from boilers, engines and GCU are measured and reported to assist the CTS in calculating delivered LNG fuel when bunkering LNG-powered vessels. Since the CTS system was a standard delivery, it was not that easy to add another calculation to the system, but the equation was known and ICAS had the data - therefore the function was implemented there.

Another example is from a test of the gas to FO changeover of the MG sets, which was performed during sea trail with full load on the generators. The test was performed by closing of GVU, and ended in a total blackout, which was not anticipated. It turned out that there was a time delay between gas supply loss and automatic changeover to FO in the control system of the MG sets, which caused the problem. The solution was for the ICAS to send a notification to MG set control system as soon as gas supply was turned off, making the changeover to FO instant, thus avoiding a total blackout. HMA implemented the function right away system was retested and approved.

7. CONCLUSION

We have now shared some of our insight when it comes to automation system design, development, work methods and interface to other systems. When applying automation to new segments like small-scale LNG, an integrated approach will technically beneficial to get the system, which perform in a way, anticipated by the operator. This new business segment is dependent on cheaper solutions to thrive and therefore it will be beneficial to use industrial standards at a high degree to minimize development costs and to make the systems future proof.

We have seen that an ever-increasing pace in the market leaves little time to specify the details when it comes to automation integration philosophy and therefore a high degree of selectiveness of vendor based on previous deliveries is applied, instead of having a general specification for more suppliers to compete against each other.

It is important to remember that even the biggest suppliers, with the most integrated systems, are basically a conglomerate of small vendors competing under the same logo.

By splitting up the control packages and defining interfaces between different systems, the control strategy becomes clearer and the risk is decreased. This philosophy is also used internally in some companies to increase portability of code and section the project between different software engineers specialising in their domain. Høglund is applying this technique to get the best possible SW product delivered to market. By focusing solely on SW, and letting others develop HW, both disciplines gain and the product becomes better than the two disciplines alone.

8. AUTHOR BIOGRAPHIES

Bengt-Olav Berntsen holds the position of R&D manager at Høglund Marine Automation. He is responsible for the R&D activities in the company. His previous experience has been maintenance and operation of marine automation systems. He has a Bachelor Degree in Automation from NTNU in Norway with several years of hands-on experience on board a variety of ships. He has a pragmatic approach with a focus on designing advanced systems with high level of user friendliness and operational qualities. This background is now used to design the automation systems of the future, which can deal with new fuel systems and technologies as well as changes in rules and regulations.

Kåre Høglund is the chairman and Business Development Manager at Høglund Marine Automation. He is responsible for new business developments of the company and by being an active company owner, pushing development. His previous experience includes development, maintenance, and commissioning of marine automation. He has a marine engineering degree from Trondheim Maritime Høgskole, Norway. His focus has always been SW development and the extensive use of thoroughly tested hardware components. He is a keen believer of remote development work and is extensively using SW simulators when developing technically challenging applications.

JETTYLESS SHIP TO SHORE TRANSFER AND BUNKERING OF LNG

J D Strachan, Houlder Ltd, UK

SUMMARY

This paper introduces the concept of a floating transfer terminal that has been developed by Houlder in association Wärtsilä Klaw LNG and Trelleborg. The paper discusses the features of the various components incorporated within the FTT system. The operation of the system is also explained. The benefits of the system when compared with a traditional shoreside terminal are also discussed.

NOMENCLATURE

Symbol	Definition
LNG	Liquefied Natural Gas
LNGC	Liquefied Natural Gas Carrier
FTT	Floating Transfer Terminal
KHOBRA	Klaw/Houlder's LNG transfer system
ESD	Emergency Shut Down
ERS	Emergency Release System
MBM	Multi-Buoy Moorings
FSRU	Floating Storage and Regasification Unit
BOG	Boil Off Gas
CAPEX	Capital Expenditure

1. HOULDER'S RECENT LNG EXPERIENCE

Houlder have been developing LNG components with our partners Klaw LNG, Wärtsilä and Trelleborg. This paper describes the Floating transfer terminal which has been developed as part of that collaboration.

We have also been heavily involved with the design of a number of significant LNG projects in recent years, having recently completed the basic design package for the two CMAL ferries for Ferguson Marine Engineering Limited. In addition, Houlder is currently contracted by Siem Shipping to provide plan approval services during the design phase of their new LNG Dual Fuel Pure Car and Truck Carrier (PCTC) vessels. The two ships are being constructed at the Xiamen Shipbuilding yard in China. We have also recently completed plan approval service on an LNG FRSU.

2. INTRODUCTION

The accessibility of LNG for small to mid-scale applications is becoming increasingly important to the growth strategies of archipelagos and coastal communities across the world. Such communities, however have limited appetite for major civil works required for new harbours, quaysides and jetties due to environmental sensitivities and lack of readily available resources and investment. Houlder have approached the problem of facilitating gas to power capability by bringing the jetty to the LNG vessel, rather than having the LNG vessel come to the jetty.

Houlder's Floating Transfer Terminal (FTT) is a flexible, rapidly deployable solution that, thanks to a low draft barge and floating hoses, can operate in the widest possible range of locations at a fraction of the cost of fixed infrastructure

3. ECONOMIC DRIVERS

3.1 REDUCED CAPITAL EXPENDITURE

A floating solution reduces the need for major civil engineering works that would be required for a jetty, mooring berth structure and breakwater, which can significantly reduce CAPEX.

The relative costs of the FTT in comparison with the Jetty is clearly dependant on the local conditions, water depth met ocean and bathymetric conditions. The studies that we have undertaken indicate that the FTT will be competitive in a number of cases. This is illustrated in Figure 1.

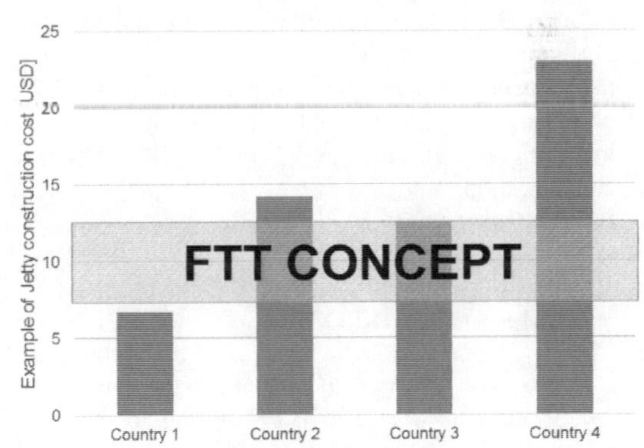

Figure 1: Cost comparison of Jetty versus FTT

The LNG FTT is built, outfitted and commissioned off-site in parallel with relatively limited civil engineering activity, reducing the overall timescale from contract award to first gas.

As self-contained mobile unit, a floating barge can be readily adapted for future and alternative deployment in the event of local changes or a desire to move location entirely. Individual components can be up or downscaled depending on requirements.

4. KEY DRIVERS

The LNG FTT solution offers the several key benefits, particularly in comparison with a fixed jetty solution, these include the following: –

4.1 VISUAL/ENVIRONMENTAL IMPACT

The visual impact of the civil engineering works involved in a typical LNG terminal comprising of a long jetty along with mooring dolphins is considerable. When coupled with the requirement to undertake dredging operations to accommodate the draught of the LNGC vessel, it is evident that the environmental impact can also be considerable.

In comparison the visual and environmental impact of the FTT barge is considerably less as the barge is only deployed when operations are underway, maintaining the quality of the local environment and minimising the risk of planning objections. This also has the added benefit of speeding up the planning process.

Additionally, it is worth noting that the barge configuration provides the minimum draught solution, thus allowing the barge to be installed in locations where water depth is limited thus further limiting requirement for dredging.

4.2 MOBILITY

A floating solution has the added benefit of allowing for:

- refuge to be sought in safe harbour during storms or hurricanes
- maintenance of the underwater components of the barge can be undertaken at local shipyard
- integration with a variety of LNGC mooring configurations
- flexibility to support future alternative applications

4.3 SELF-CONTAINED UNIT

A complete LNG barge system can be built cost effectively and transported as a single unit to site with limited integration requirements. The configuration of the vessel as a simple barge has the additional benefit of ensuring that construction is straightforward and can therefore be undertaken in a wide variety of shipyards, including local shipyards.

4.4 VESSEL COMPATIBILITY

Due to simplicity of its interfaces, the LNG FTT can be used for export and import of LNG from a wide range of LNG tankers and LNG fuelled vessels. The system fits the vessel, rather than asking the vessel to fit established, fixed infrastructure of jetty and mooring dolphins.

4.5 SHORT CONSTRUCTION TIME

The LNG FTT is built, outfitted and commissioned off-site in a local shipyard in parallel with the relatively limited civil engineering activity, reducing the overall timescale from contract award to first gas.

4.6 FUTURE DEPLOYMENT

As a self-contained mobile unit, the floating barge can be readily adapted for future and alternative deployment in the event of local changes or a desire to move location entirely. Individual components can be up or downscaled depending on requirements.

5. SYSTEM OVERVIEW AND MAIN COMPONENTS

The FTT is made up of;

- Self-propelled or towed barge
- Cryoline floating hose
- Vertical or horizontal axis reel to store the hoses and umbilical. This can be positioned on the barge or on the shore.
- KHOBRA LNG transfer system on the barge

Figure 2: Floating Transfer Terminal

5.1 FTT BARGE

An efficiently designed steel or concrete barge manoeuvres the floating transfer hoses and carries the KHOBRA that safely and securely makes the connection to the delivery vessel moored in the vicinity of the facility. The barge is sized such that stability is guaranteed and motions are reduced to an acceptable level whilst keeping costs and visual impact low.
The barge can either be self-propelled or towed into position by a separate tug, depending on specific operational requirements.

The barge typically accommodates: a control station; nitrogen accumulators for gas process operation; hydraulic accumulators and Hydraulic Power Unit; and storage for loose equipment. These components can either be powered via the umbilical cable or using small generator fitted on board.

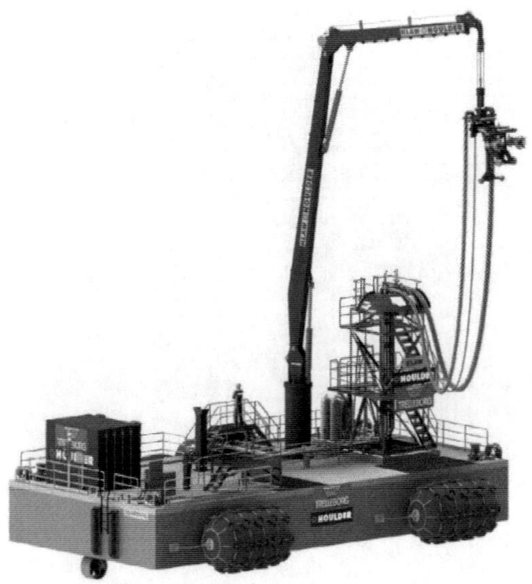

Figure 3: The FTT Barge

5.2 SEAKEEPING AND BARGE MOTIONS

Whilst it is clear that the motions of a semisubmersible configuration would be lower than that of a conventional barge the barge configuration is still considered to be well suited to this role. The FTT barge will operate in and be moored alongside the lee of the LNG tanker or receiving vessel, therefore it is relatively sheltered from waves with shorter periods. The FTT barge is considerably smaller than the LNG carrier/receiving vessel and as such the motions of the barge will follow those of the LNG carrier/receiving vessel in waves of longer periods such as swells. Therefore, although the motions of the FTT barge are likely to be larger than those of a semisubmersible the relative motions to those of the ship are likely to be similar.

5.3 FLOATING CRYOGENIC HOSES

Trelleborg Cryoline Floating Hoses are used for the transfer of LNG and Boil Off Gas (BOG) between the shore and the barge. This proven technology provides high thermal insulation and an integrated leak monitoring system to achieve safe transfer with minimum BOG generation.

The Cryoline hoses have a maximum allowable operating pressure of 20 bar and can be supplied with 6" to 20" diameters depending on the specific transfer requirements.

The Cryoline hoses are delivered in 12m sections, which allows for additional lengths to be added if required.

The liquid and vapour return hoses are terminated at the barge with protective bend restrictors to ensure the minimum bending radius is maintained. Cryogenic piping then leads to the KHOBRA System where the connection to the ship's manifold is made.

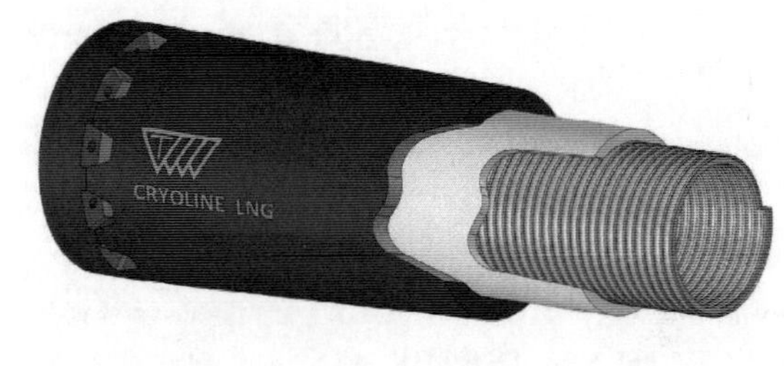

Figure 4: Trelleborg Cryoline Hoses

5.4 SERVICE UMBILICAL

A buoyant service umbilical is deployed with the floating hoses that include a power cable, Ship to Shore ESD link, communications, control lines and a nitrogen supply line.

5.5 REELS

Powered reels are employed shore-side to deploy, recover and store the floating hoses and service umbilical.

Either vertical or horizontal axis reels can be used depending on Client preference or operational requirements.

As an alternative the reels may be located on the FTT barge to provide increased flexibility of operation.

Figure 5: Vertical Reel for storage of Cryoline hose

5.6 KHOBRA

The FTT employs a novel hybrid LNG transfer system that provides a safe and efficient hybrid hose handling solution which is well suited to this application. Houlder & KLAW LNG's transfer system facilitates the transfer of LNG from the LNG carrier to the FTT, or from shore installation to a LNG fuelled vessel, in a safe and controlled way. The system consists of a deployment crane, a motion compensated deployable Emergency Release System (ERS) manifold and Cargo System. The KHOBRA system typically includes:

- knuckle boom crane
- Two transfer lines – one liquid and one vapour return
- Safety Breakaway couplers (Active ERS valves)
- Quick connection coupler
- Emergency Shut Down (ESD) Link – SIL 2 Rated

Key to the success of the KHOBRA transfer system is a tensioner and compensation system known as the Deployable ERS (DERS) that monitors and compensates for FTT/LNGC/receiving vessel motions and changes in draught. It protects the alongside vessel's manifold from excessive forces during transfer and monitors the manifold position to check for vessel drift and trigger emergency shutdown or, ultimately, release through breakaway couplers if the limits are reached.

5.7 DEPLOYABLE ERS

KHOBRA's deployable ERS offers secure connection to a wide range of vessel manifold configurations and facilitates emergency shut down and release.

The specially designed frame eliminates loads at the manifold connection point, removing the risk of excessive loads being transmitted inadvertently due to movement of the vessel. The system is also equipped with cryogenic dry coupler for safe and quick connection and disconnection. The system as designed fully supports ESD1 and active ESD2 functions.

The Deployable ERS has the following components:

- TCS – Tension Compensating System to maintain a constant tension.
- Accumulator – Stored hydraulic power that removes the DERS dependency on an HPU during operation.
- Control Umbilical – Provides the control signal for the breakaway ERS and ESD link.
- Transfer Hoses – These carry the LNG and vapour return
- ERS – Upon detection of a potential hazard the hydraulically powered ERS safely detaches the LNG supply from the receiving manifold
- Connection Hose – used to connect with the receiving vessel manifold
- QCDC – Quick Connect Disconnect Coupler
- Landing Frame - Eliminates loads being transmitted to the receiving manifold connection point.

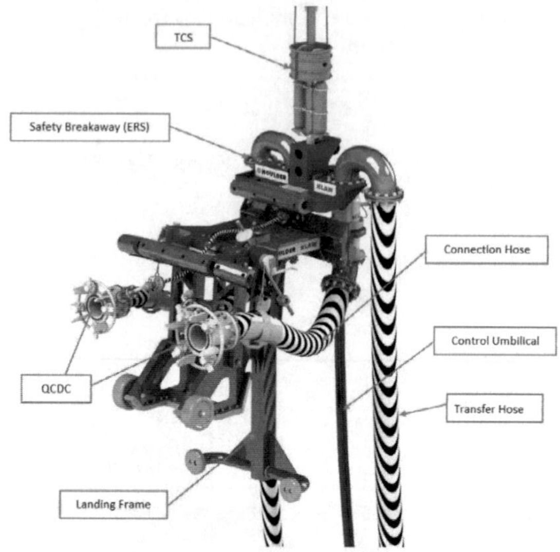

Figure 6: Components of the Deployable ERS (DERS)

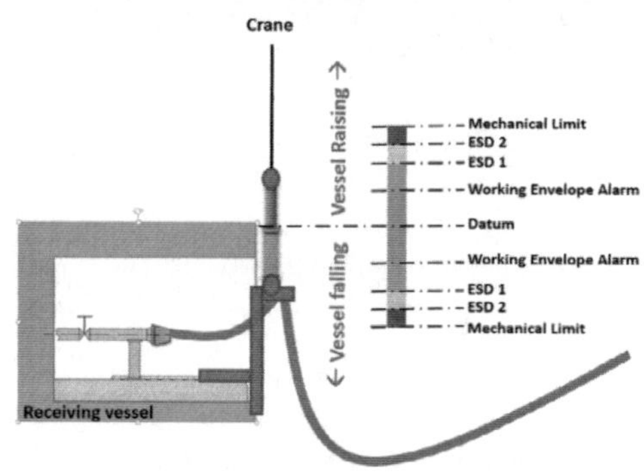

Figure 7: Envelope limit for the DERS

5.8 KHOBRA ENVELOPE LIMIT DETECTION

A tensioner control system allows the FTT barge and the alongside vessel to move independently of each other within a safe operating window. When relative motions exceed the predetermined safe limit, the ESD procedure is automatically initiated eliminating all possible hazards.

The ESD can also be initiated manually by the operator.

- ESD 1: Emergency Shutdown Stage 1 – An alarm is activated, and the LNG supply operation is shut down in a quick controlled manner by stopping the transfer pumps and closing the shutdown valves and other relevant equipment in ship, FTT and shore systems.

- ESD 2: Emergency Shutdown Stage 2 - shuts down the transfer operation (ESD 1) and uncouples the transfer system after closure of ERS isolation valve/s.

5.9 KHOBRA SYSTEM

The System comprises two transfer lines, each includes:

- Composite flexible cryogenic transfer hose
- Safety Breakaway couplers (Hydraulically powered active ERS valve)
- Dry Disconnect Coupler

During transfer operations, the cryogenic hoses are to be connected to the receiving/replenishment vessel LNG and vapour manifolds.

A Nitrogen purging line is connected and disconnected to the transfer lines for purging operation as required.

6. SAFETY

The effective operation of an emergency shutdown ESD system is essential to ensure safe transfer operations. The FTT benefits from using the KHOBRA's Incorporates a hydraulically powered "active" Emergency Release System (ERS).

The system monitors the relative position of the ship's manifold to check for vessel drift and can trigger emergency shutdown and ultimately initiate the release of the breakaway coupling if those limits are reached.

The Safety Control System fully supports ESD1 and ESD2 functionality connected to vessel drift monitoring system. Designed to comply with the highly demanding SIL2 control safety requirement, once the ESD2 signal is triggered, the KHOBRA performs an emergency disconnection, retrieving the transfer hoses from the receiving vessel.

Nitrogen is used on the FTT for purging and inerting hoses and connection flanges during the connection and disconnection phases the operation.

A dry chemical powder firefighting system is incorporated within the FTT design to protect LNG systems on-board.

The FTT barge and LNG systems will be designed fabricated and installed in accordance with the relevant standards as defined below.

6.1 APPLICABLE STANDARDS

A number of international standards are expected to be applicable to different components of the FTT:-

Barge

- Class rules for steel barges and pontoon

Reels

- CE marking and Class rules for offshore / marine structure

Cryoline Hoses

- EN1474/2 Installation and equipment for liquefied natural gas. Design and testing of marine transfer systems. Design and testing of transfer hoses

Transfer system

- EN1474/2 Installation and equipment for liquefied natural gas. Design and testing of marine transfer systems. Design and testing of transfer hoses
- EN1474/3 Installation and equipment for liquefied natural gas. Design and testing of marine transfer systems. Offshore transfer systems
- ISO 16904 Petroleum and natural gas industries -- Design and testing of LNG marine transfer arms for conventional onshore terminals
- ISO 20519:2017 Ships and marine technology -- Specification for bunkering of liquefied natural gas fuelled vessels
- ISO 16904 Petroleum and natural gas industries -- Design and testing of LNG marine transfer arms for conventional onshore terminals
- ISO 20519:2017 Ships and marine technology -- Specification for bunkering of liquefied natural gas fuelled vessels
- ISO-TS 18683 2015 - Guidelines for Systems and Installation for Supply of LNG as Fuel to Ships
- USCG /Port and Local Regulations
- Ship Owner Requirements
- SIGTTO guidelines
- OCIMF guidelines

6.2 RISK ASSESSMENT

With the risk assessments to be undertaken in accordance with:-

- PD ISO-TS 16901 2015 - Guidance on Performing Risk Assessment in the Design of Onshore LNG Installations including the Ship-Shore Interface

6.3 TESTING AND COMMISSIONING

The testing and installation of the system will be in accordance with:-

EN1474/3 Installation and equipment for liquefied natural gas. Design and testing of marine transfer systems. Offshore transfer systems

7. OPERATION

The function of the FTT is best explained within the context of how it operates alongside the LNG vessel. Operational processes can be divided into the following stages:-

7.1 VESSEL MOORING

The FTT is designed for use in shallow waters where there is no fixed jetty or available jetty space for the LNGC to berth alongside.

A Multi-Buoy Mooring will, therefore, be utilised for mooring the LNGC and holding it on a fixed heading for the FTT to berth alongside for the transfer of LNG from the vessel to shore.

As it is expected that in many cases the FTT will be utilized in positions where the existing marine infrastructure is minimal, therefore tugs will probably not be available for assisting in the berthing operations.

The siting of the MBM will need to be agreed with the relevant local authorities and care taken to avoid damage to the environment and subsea cables and pipelines etc.

Consideration for the siting of the MBM and mooring configuration should include:

- Maximum wind velocities under which operations may be carried out.
- Wind direction.
- Current.
- Tidal variations.
- Maximum wave heights and swell.
- Possibility of passing ships causing surging.
- Draft changes during cargo transfer.

There are two options for mooring the LNGC/receiving vessel.

7.2 MOORING OPERATION – VESSEL TO MBM – USING ANCHORS FOR "RUNNING MOOR"

The basic layout for this type of berth will have buoys to moor the aft of the vessel to, each buoy being independently anchored to the seabed by a pendant chain and anchor system.

The buoys will be equipped with quick release hooks, these should have an SWL which is not less than the MBL of the largest rope anticipated to be used by the vessels.

The vessel will moor in position by carrying out a "running moor", this operation is normally undertaken without tug assistance and will require the full and efficient use of the ship's mooring equipment and the expertise in ship handling of the crew.

The standard mooring equipment of the vessel will be adequate for this operation, all mooring lines to be from the vessel. Two lines may be required to be run out to some, or all, of the buoys.

7.3 PROCESS FOR MOORING AT BUOYS

The mooring operation will start with the vessel performing a "running moor", with the vessel proceeding slowly towards the forward end of the berth, almost at 90° to the final position. At the correct moment the starboard anchor is let go and the chain run out as the vessel continues moving ahead with the engine going astern. Once the vessel is stopped in the water (this may be ascertained by observing the vessel's wake) the port anchor is now let go. The engines and helm are now used in conjunction with paying out the port anchor whilst heaving in the starboard anchor to get the stern to swing clear of the nearest buoy and the vessel moving astern into the sector between the buoys. Mooring lines may now be sent, using the line boat, to the buoys and when secure these are used to help in controlling the swing and pulling the vessel astern into the berth.

7.4 LNG FTT DEPLOYMENT

The FTT barge is harboured at the shore allowing for quick mobilisation and deployment once the vessel is in position. It will manoeuvre towards the offloading/receiving vessel under its own propulsion and moor itself alongside. The flexible hoses and service umbilical deploy from their reels as the barge manoeuvres.

Figure 8: FTT Barge connected to LNGC

7.5 KHOBRA CONNECTION AND DISCONNECTION

Upon completion of mooring and establishing a safety communication system between the FTT barge and the receiving/unloading vessel, the KHOBRA Deployable Emergency Release System (DERS) will be deployed. The DERS is manoeuvred into position by personnel on the barge using the crane, establishing a connection between the transfer hoses and the vessel manifolds.

The design and construction of the KHOBRA DERS aims to minimise the risk of: operator errors, injury to persons, damage to assets, and methane venting during transfer operations.

7.6 EMERGENCY SHUT DOWN (ESD)

A tensioner control system allows the receiving/replenishment vessel and FTT barge to move independently of each other within a safe operating window. When relative motions exceed the predetermined safe limit, the ESD procedure is automatically initiated eliminating all possible hazards. Manual ESD activation is also possible.

7.7 CARGO OPERATIONS

The LNG FTT process system has been developed following the principles of a traditional loading and unloading terminal. It includes features and design principles that facilitate current transfer operations.

Key design features include;

- Presence of on-board nitrogen reservoir and additional shore nitrogen supply through the service umbilical for purging operation.

8. FTT DEVELOPMENT PROGRAMME

Houlder are continuing to develop the FTT with our partners. A phased/gated and risk mitigation approach has been chosen to allow the project stakeholders to have continuous and effective monitoring of the project's progress and permit a fluent and proactive interaction between the vendor and key user.

Four principal phases form the development and supply programme

- Phase 1: Conceptual study
- Phase 2: Detailed Engineering
- Phase 3: Execution, Commissioning and Start-up.
- Phase 4: Handover and Operations.

9. CONCLUSION

The FTT has been developed in order to reduce the capital expenditure required to install a small to mid-scale LNG terminal.

The system incorporates advanced technologies within the floating cryogenic hoses and innovative LNG transfer system.

The FTT system as designed has a number of benefits that make it well-suited to small/mid-scale LNG transfer for incorporation within gas powered power stations as developed by Wärtsilä.

10. ACKNOLEDGEMENTS

The author would like to express his gratitude to his colleagues within Houlder and Wärtsilä, Trelleborg and Klaw LNG for their cooperation and assistance in developing the product and this paper.

SLOSHING AND SWIRLING IN MEMBRANE LNG TANKS AND THEIR COUPLING EFFECTS WITH SHIP MOTION

M Arai and **G M Karuka**, Yokohama National University, Japan, **H Ando,** Monohakobi Technology Institute, Japan

SUMMARY

A model experiment was carried out using an LNG carrier model with partially filled tanks and the results were compared with numerical simulations. The ship motion is calculated using the strip-method and the liquid flow computation is based on the finite difference method. The coupled behaviour between the ship motion and the partially filled tanks is studied using the measured ship motion tank horizontal forces. The tanks have a length to breadth ration near one, and because of that sloshing and swirling motions were observed during the tests. The influence of these difference modes on the coupled ship motion is discussed.

1. INTRODUCTION

When LNG carriers or Floating LNG operate with partially filled tanks, there is a possibility of violent liquid motions occurrence inside the tanks. This phenomenon is known as sloshing. Sloshing can compromise the tanks structural integrity and in order to avoid it, ships are only allowed to operate with nearly full or nearly empty conditions. In the future, however, partially loaded condition might be necessary in some operations such as LNG offloading, usage of LNG in fuel tanks, and flexibility increase on LNG logistics.

The liquid motion inside the tank may interact with the ship motion, resulting in a coupled motion. In Nam *et al.* (2009) work, the ship motion is calculated with the impulse response function (IRF) method, and the non-linear sloshing flow is calculated with a FDM method. Regular flow was tested, and the RAO was compared with an experiment. It was found that the sloshing induced loads are not linear in respect to the motion amplitude.

Jiang *et al.* (2015) used a similar method, but using the Volume of fluid Method interface capturing was used for solving the internal flow, and it was concluded that the coupling effects are especially larger for lower filling levels.

Hu *et al.* (2017) studied the sloshing effect on a FLNG system, using a potential method to calculate both the ship motion and the internal flow. According to the results, the influence on pitch, yaw and heave were negligible while a noticeable effect was observed near the tank natural frequency for sway and surge.

In the present work, a model experiment was carried using an LNG carrier model with a 1:68.75 scale including forward speed, and the results were compared with numerical simulations. The ship motion is calculated using the strip-method and the liquid flow computation is based on the finite difference method. The coupled behaviour between the tank motion and the partially filled tanks is studied using the data obtained from the ship motion sensors and the bi-directional load cell installed under the partially filled tank.

The tanks used in the experiment have a breadth to length ratio near one, and sloshing and swirling motions were observed during the tests.

When a partially filled tank has a length to breadth ratio near 1, there is a chance that, in addition to sloshing, a rotational flow, i.e. swirling, may also occur.

After analysing the tank dimension and arrangements of a number of LNG carriers, it was noted that some of the designs have tanks with length to breadth ratio near one. Tanks located near the aft and forward in special, tend to be shorter in order to accommodate to the hull geometry. Therefore, the problem is worth being investigated.

Swirling is a common phenomenon for spherical tanks, as described in described in Faltinsen & Timokha (2013) and Arai *et al.* (2016), for example. Swirling in prismatic tanks is mentioned when tanks are excited in irregular motion in Chen *et al.* (2011). Karuka *et al.* (2017) examined the effects of swirling for regular sway excitation, identifying the conditions in which the phenomena may occur. It was found that the swirling maximum impact pressure is lower than sloshing impact pressure, but can still be significant, and the affected areas in the tank are different.

In general, there is no information available concerning the effects of swirling in coupled ship motion. Therefore, the influence of sloshing and swirling in the coupled ship motion is discussed in this paper.

2. NUMERICAL METHOD

The coupled system includes two distinct problems: the ship-motion problem and the sloshing problem. figure 1 shows a flow chart of the computation for solving the coupled ship motion and sloshing problems

2.1 SHIP-MOTION PROBLEM

The ship motion is calculated by using the strip-theory as show in equation 1.

$$(\{m_{ij}\} + \{a_{ij}\})\{\ddot{\varsigma}_j\} + \{b_{ij}\}\{\dot{\varsigma}_j\} + \{c_{ij}\}\{\varsigma_j\} = \{f_i\}_{wave} + \{f_i\}_{slosh} \qquad (1)$$

Where $\varsigma_j = ship\ motions$ (1-surge,2-sway,3-heave,4-roll,5-pitch,6-yaw); m_{ij}= mass or moment of inertia; a_{ij} = added mass; b_{ij} = damping coefficients; c_{ij} = restoring coefficients; f_{wave} =wave exciting force.

- m_{ij} is the mass of the ship without liquid
- The center of gravity is calculated for the ship without liquid
- For the restoring coefficient of rolling c_{44}
- $c_{44} = m'g.GM$, where GM is the metacentric height for the ship hull without liquid, m' is the total mass including the ship and the liquid

The hydrodynamics forces and coefficients were calculated using the Strip Method. At each section of the ship, the two-dimensional radiation and diffraction problems were solved in the frequency domain by using the Boundary Element Method.

Because of the nonlinearity of the sloshing flow, the simulations are carried out in time domain. The Newmark-beta method has been employed to solve equation 1 in the time domain.

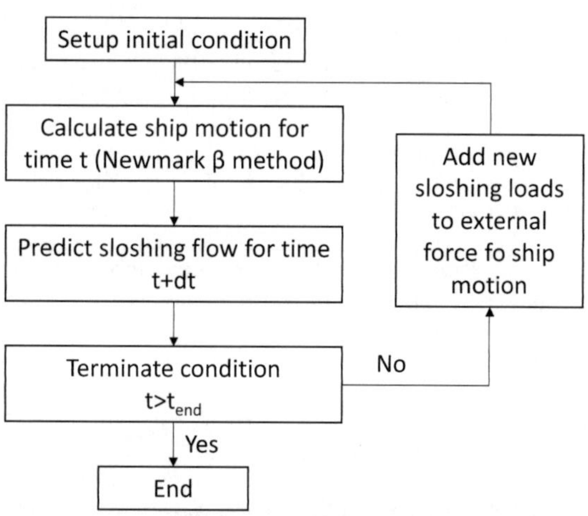

Figure 1: coupling program flowchart.

2.2 THE SLOSHING PROBLEM

The numerical method used in the analysis of sloshing was based on the finite difference technique. The numerical method is outlined below.

2.2 (a) Governing Equations

A coordinate system *o-xyz* fixed to the moving tank was adopted. Assuming an incompressible and inviscid fluid, the equations governing the liquid cargo motion inside the tank are the mass continuity equation

$$\frac{\partial u}{\partial x} + \frac{\partial v}{\partial y} + \frac{\partial w}{\partial z} = 0 \qquad (2)$$

and Euler's equations of motion:

$$\begin{aligned}
\frac{\partial u}{\partial t} + u\frac{\partial u}{\partial x} + v\frac{\partial u}{\partial y} + w\frac{\partial u}{\partial z} &= -\frac{\partial p}{\rho \partial x} + \frac{f_x}{\rho} \\
\frac{\partial v}{\partial t} + u\frac{\partial v}{\partial x} + v\frac{\partial v}{\partial y} + w\frac{\partial v}{\partial z} &= -\frac{\partial p}{\rho \partial y} + \frac{f_y}{\rho} \\
\frac{\partial w}{\partial t} + u\frac{\partial w}{\partial x} + v\frac{\partial w}{\partial y} + w\frac{\partial w}{\partial z} &= -\frac{\partial p}{\rho \partial z} + \frac{f_z}{\rho}
\end{aligned} \qquad (3)$$

Where

u, v, w: velocity components with respect to the coordinates fixed to the tank;

f_x, f_y and f_z: x, y and z components of external forces,

p: pressure,

ρ: density of the liquid.

The exciting motion of the tank is taken into account by the external forces f_x, f_y and f_z in the motion equations.

2.2 (b) Finite Difference Approximation

To reduce the computational time and to simplify the numerical method, a staggered mesh system with constant grid spacing Δx, Δy and Δz in the x, y and z directions was used. The variables u, v and w were evaluated at the cell faces, while p was evaluated at the center of the grid cells. Once the initial conditions of the problem were applied, the velocities u, v and w were estimated for the next time step by using the motion equations. The variables u, v, w and p were iteratively adjusted to satisfy the continuity equation and boundary conditions. Details can be found in Arai *et al.* (2002) and Cheng *et al.* (2005).

2.2 (c) Rigid wall

Rigid boundaries are modelled by setting zero normal velocity on the wall. Free-slip with the assumption of inviscid was used.

2.2 (d) Free Surface

The position of the free surface was evaluated by using a height function h, whose value was updated at every time step by applying the kinematic condition:

$$\frac{\partial h}{\partial t} + u\frac{\partial h}{\partial x} + v\frac{\partial h}{\partial y} = w \qquad (4)$$

where $h=h(t,x,y)$ is the height of the free surface measured from the tank bottom and is a function of t, x and y. The atmospheric pressure p_{atm} is set at the free surface location.

2.2 (e) Impact on tank ceiling

To achieve a stable assessment of the impact pressure at the tank ceiling, a transition of the boundary condition from a free surface to a rigid wall proposed by Arai *et al.* (2002) is considered. A detailed explanation of this condition can also be found in Cheng *et al.* (2005).

3. EXPERIMENTAL SETUP

3.1 INTRODUCTION

An experiment was carried out at NMRI (National Maritime Research Institute) in Japan in December 2016. The goal of the experiment was to investigate the coupled behaviour of an LNG ship model and prismatic tanks filled with liquid. The ship motion and the tank internal forces were measured.

The scale between the reference ship and the model ship is 1:68.75. The force was measure though a force load cell that could measure the force in the transverse and longitudinal directions. The load cell was installed on the forward tank. Heave, sway and surge were measured by accelerometers and roll, pitch and yaw by gyroscopes.

(a) LNG carrier model

(b) Tank model

Figure 2: Pictures of the model ship and model tank.

3.2 Ship model description

The model ship was designed bases on a reference LNG carrier. The LNG carrier was slightly modified for practical purposes and scaled to the model size. The model dimensions are described in table 1.

Table 1: Model ship properties.

MODEL MAIN DIMENSIONS		
Lpp	[m]	4.00
B	[m]	0.64
D	[m]	0.38
T	[m]	0.17
ROTATION CENTER POSITION		
x_c	[from AP,m]	1.98
y_c	[from center line,m]	0.00
z_c	[from keel,m]	0.17

3.3 TANK DESCRIPTION

In the ship used as reference, four LNG tanks are used. However, for practical purposes, water was used instead of LNG in the experiment. Since the water density is approximately twice as heavy as LNG, the number of the tanks was reduced from four to two in order to preserve the cargo weight between the real and the model scale. The tank dimensions are described in table 2 and figure 3.

Liquid and solid tests were performed to compare the liquid effect on the ship motion. The liquid mass for each tank is 35.85kg and 60.93kg for the 30% and 50% filling levels, respectively.

In table 3, the model ship properties are described in the solid condition and in the liquid condition.

Table 2: Tank properties.

TANK DIMENSIONS			
Length		[m]	0.565
Width		[m]	0.554
Depth		[m]	0.400
Bottom chamfer height		[m]	0.056
Top chamfer height		[m]	0.121
TANK CENTER POSITION		Tank 1	Tank 2
x	[from AP,m]	2.69	1.35
y	[from center line,m]	0.00	0.00
z	[from keel,m]	0.29	0.29

*the distance from the keel to t
he tank internal bottom is 0.079m

Table 3: Test conditions.

SOLID CONDITION		30%	50%
Δ	[kg]	287.310	337.470
DRAFT	[m]	0.152	0.176
DRAFT AP	[m]	0.159	0.176
DRAFT FWD	[m]	0.148	0.176
GM_t	[m]	0.147	0.120
KG	[m]	0.162	0.174
R_{xx}	[m]	0.202	0.197
R_{yy}	[m]	0.933	0.905
LIQUID CONDITION		50% case	30% case
Δ	[kg]	287.310	337.470
DRAFT	[m]	0.152	0.176
DRAFT AP	[m]	0.159	0.176
DRAFT FWD	[m]	0.148	0.176
GM_t	[m]	0.147	0.120
KG	[m]	0.162	0.174
R_{xx}	[m]	0.213	0.215
R_{yy}	[m]	1.006	1.000
R_{zz}	[m]	1.000	0.997

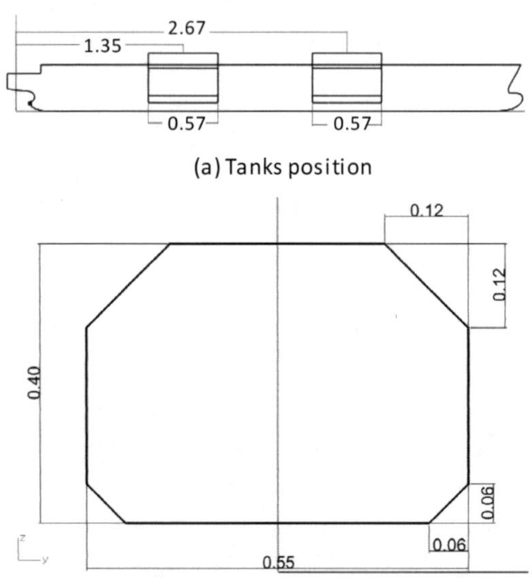

(a) Tanks position

(b) Tank dimension

Figure 3: Tank dimension and positioning

3.4 TEST CASES

Regular wave excitation was carried out including two filling levels: 30% and 50; two cargo states: solid and liquid; for a 90deg incident wave. For all cases, the ship had an advance speed of 1.117m/s. The incident wave amplitude is constant at 4cm.The first mode of the natural frequency f_1 was estimated by a linear theory suggested by Abramson *et al.* (1974) for rectangular shaped tanks in equation 4. The estimations are shown in table 4.

$$f_1 \approx \frac{\sqrt{\left(g\pi/B_t\right)\tanh\left(\pi h/B_t\right)}}{2\pi} \tag{5}$$

Table 4: natural periods estimation

Filling level	Tank Transverse T_1	Tank Longitudinal T_1
	[s]	[s]
30%	1.094	1.113
50%	0.934	0.948

4. RESULTS

4.1 REGULAR BEAM WAVES

According to figure 4 in both 30% and 50% filling conditions, the roll natural period of the ship becomes longer with the liquid cargo when compared to the solid cargo case. The T_1 of the ship goes from 1.308s to 1.613s (an increase of 0.305s) in the 50% case and from T_1=1.215s to T_1=1.616s (an increase of 0.398s). This happens due to the free-surface effect. The presence of the tanks, means that the ship has lower area restoring moment of inertia in the longitudinal direction, and hence, a longer natural period in roll. In the 30% condition, the maximum roll amplitude at the ship natural frequency is significant lower than the maximum roll amplitude in solid condition. For the 50% condition, the maximum roll amplitude is similar for both liquid and solid cases.

For sway, a local peak is observed near the tank transverse natural period. The reason for the peak is that near the tank

natural period there is a 180 degrees phase inversion between the ship and the liquid motion. The local maximum corresponds to the period where the ship motion is in phase with the liquid motion. The local minimum corresponds to the period when the ship motion and the liquid motion are out of phase.

When the tank is excited in its natural frequency, the phase difference between the excitation and the liquid motion is 90deg.

For heave, no significant change is observed between the liquid and solid cargo cases, which is consistent with the results in Hu *et al.* (2017). Results for pitch and yaw were omitted because since the excitation is lateral, almost no response is observed.

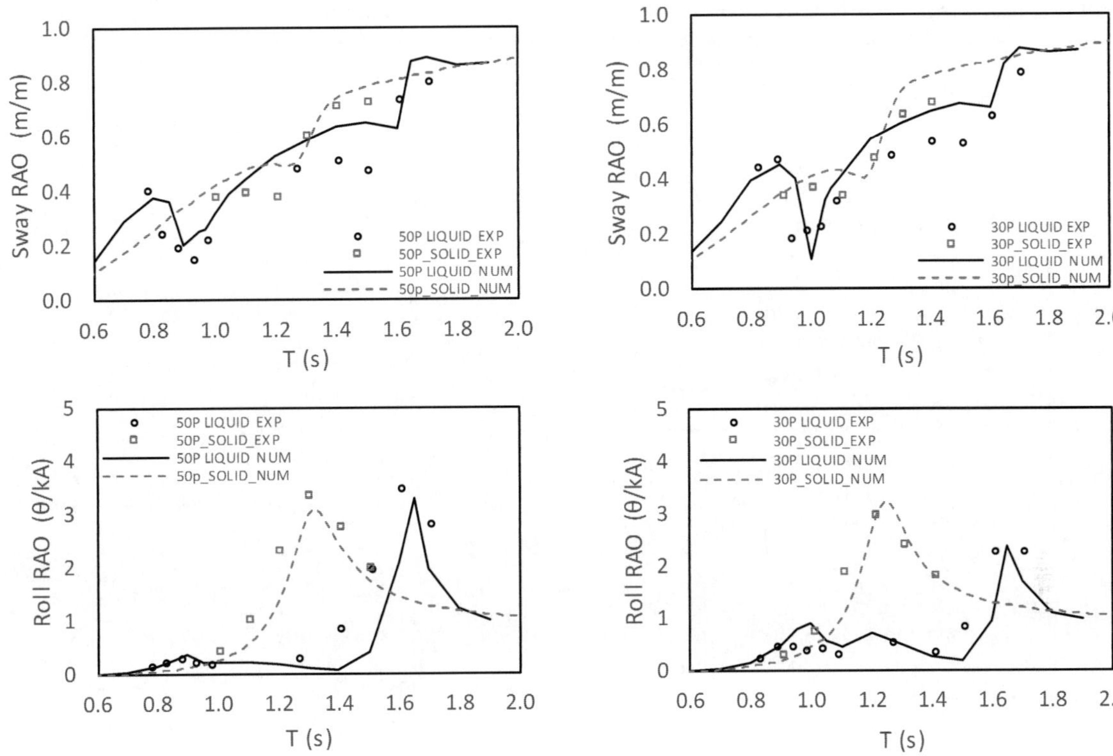

Figure 4 – RAO for 50% (left) and 30% (right) filling levels, beam sea

In general, the results are very similar between 30% and 50%. However, in the 30% the influence of the liquid cargo in the roll and sway motions are clearly larger than in the 50% case. The reason is that the system the 30% and 50% conditions have approximately a 15% difference in displacement, but the internal tank force is of same magnitude near the tank natural frequency. In addition, in low filling levels, the free-surface has a larger deformation near the tank natural period, because non-linear effects are larger.

4.2 FORCE MEASUREMENT

The forces were measured in a load cell mounted on the forward tank. model. The forces were measured in local coordinate system, which rotates in roll, pitch and yaw with the ship.

We can see that near the tank natural period, when sloshing or swirling occurs, there is an increase in the measured forces. The large transverse force measured near the ship's natural period is due to the large roll inclination of the model. When the model is inclined, the liquid pressure on the tank is high, although the free surface moves slowly.

The forces were converted to a global coordinate system fixed parallelly to the still free-surface and then compared with numerical results. The transverse force of the slow free-surface movement in the global coordinate system is small because the force exerted to the tank is the liquid weight (vertical force), and hence the horizontal forces are small. Choosing the global coordinate system to show the results emphasize the forces generated by dynamic forces in the liquid motion rather than the forces generated by the liquid weight.

It can be seen in figure 6 that the sloshing and swirling happens near the tank natural frequency.

When transverse sloshing occurs, there is a sharp increase in the transverse force. When swirling occurs, there is an increase in both F_x and F_y.

Although the force comparison shows good agreement when dynamic effects are small, there are significant difference in the result near the tank natural period. It can be noted that the liquid effect on the ship motion can be predicted with reasonable accuracy, but the opposite is not true. The sloshing mode or swirling mode is very sensitive to the tank excitation.

In general, the comparison between F_y is reasonably predicted. However, F_x results has considerable differences. Such disparities were observed when different modes occurred. For example, in some cases in the experiment swirling occurred, while in the simulation sloshing occurred. When the tank is excited only in sway direction. The occurrence of swirling can be identified by checking the relation between F_x and F_y. When F_y is large and F_x is low, transverse sloshing is occurring. When both F_x and F_y are large, swirling is occurring.

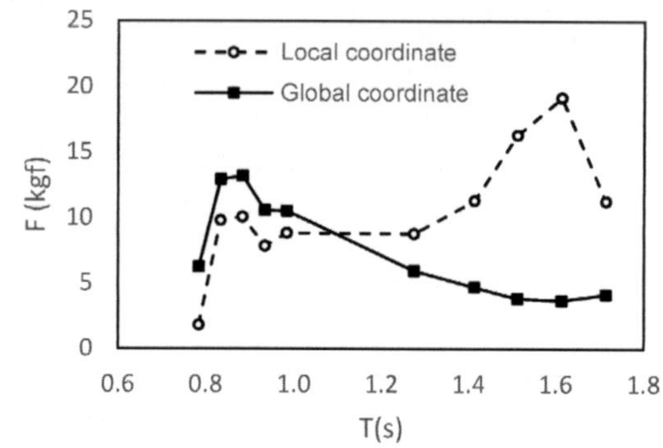

Figure 5: Forces in local and global coordinate systems

When T_l=0.934s, swirling occurs, and both F_x and F_y are excited, having almost the same magnitude and with a phase difference of 90 degrees. The phase diagram has a circle shape.

When T_l=0.984s, transverse sloshing occurs, the transverse force is predominant, and the diagram phase has the shape of a narrow ellipse.

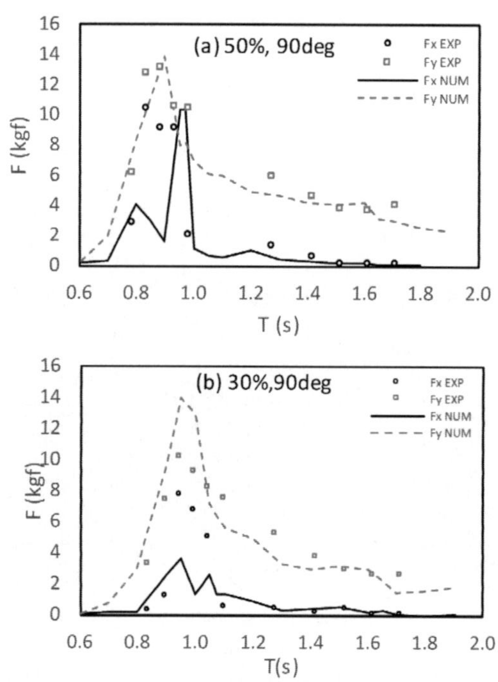

Figure 6: Measured forces in global coordinates

Figure 7 shows the cases where T= 0.934s and T=0.984s, respectively, are shown.

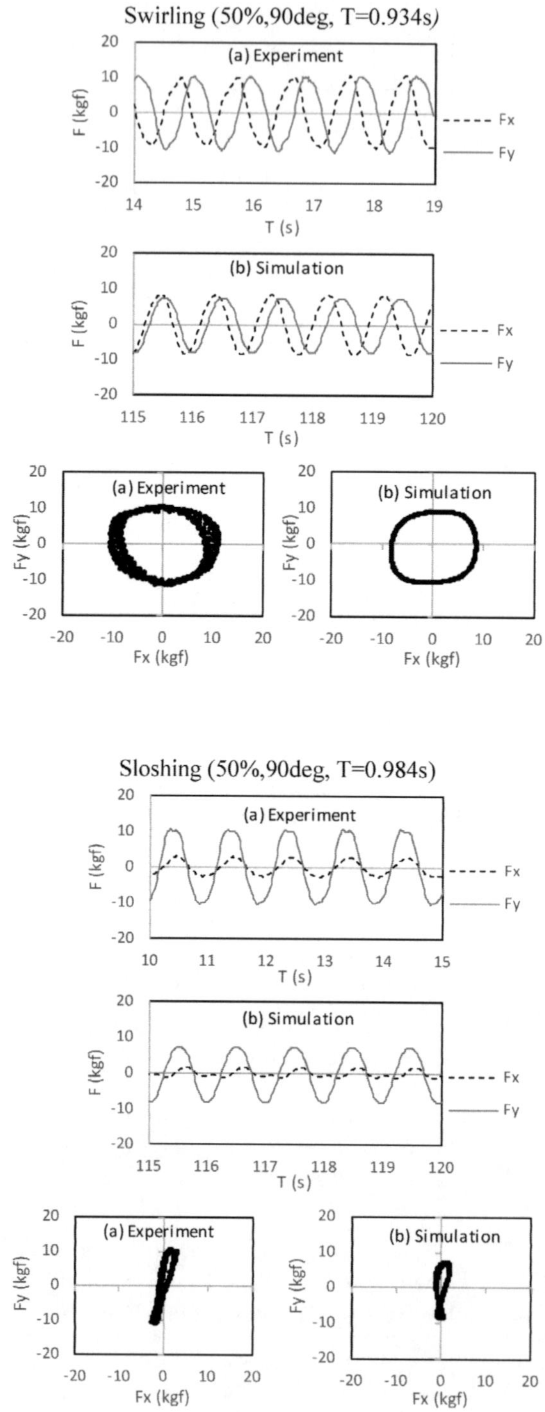

Figure 7: comparison between simulation and experiments for sloshing and swirling

4.4 INFLUENCE OF SLOSHING AND SWIRLING IN THE COUPLED SHIP MOTION

The case described in figure 7 shows swirling occurrence in the counter-clockwise direction. In the experiment, the rotation did not have a preferential direction, with both clockwise and clockwise swirling occurring. It is believed that small disturbances, such as deformations in the initial free-surface can trigger the rotation in one direction or the other.

In the simulation, it is possible in some cases to control the rotation direction by adding artificially a small inclination angle to the initial free-surface. An initial -2.0deg and a 2.0deg inclination around the y axis resulted in a counter clockwise and clockwise rotation, respectively.

In order to understand the effects of the rotation direction on the tank forces and the coupled motion, both cases were investigated. In addition, in order to compare the swirling effects to the sloshing effects, a third case where the tanks are 10% longer were also included. With a longer tank, only sloshing occurs. The liquid density of the computation liquid was adjusted to keep the liquid mass constant.

Figure 8 shows that depending on the rotation direction, the maximum force on the longitudinal direction is different. As expected, the longitudinal force for the transverse sloshing case is minimum. The transverse force F_y have nearly the same magnitude and phase for all the three cases.

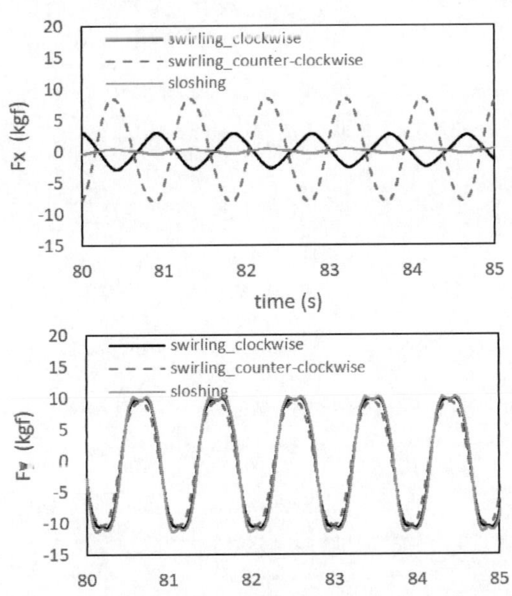

Figure 8: tank forces for different sloshing and swirling for 50% filling level, beam wave, T=0.934s.

In terms of coupled ship motion, the yaw and pitch motions are very small for beam waves and the heave has almost no influence from the liquid motion, so only sway and roll are shown in figure 9. Not a large variation is observed, but in terms of sway, the sloshing case is about 20% lower than both the swirling cases, whereas in terms of roll, the swirling in counter-clockwise is about 9% lower than sloshing and the clockwise swirling.

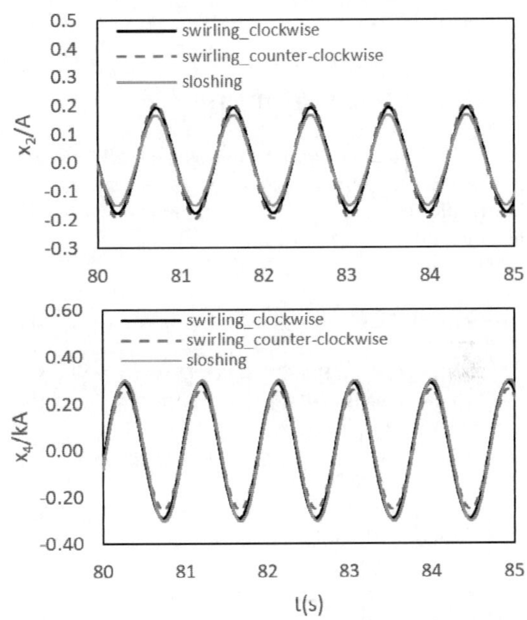

Figure 9: effect of sloshing and swirling on the coupled ship motion for 50% filling level, beam wave, T=0.934s.

5. CONCLUSIONS

A numerical method was developed to estimate the coupled behaviour of a ship and its internal tanks. A reasonably accurate prediction of the ship motion was obtained. The results are consistent with the results from different experiments and researches. The main effects of the liquid motion in the ship motion are: 1. The roll natural period becomes longer due to the free-surface effect. 2 A local roll peak occurs near the tank natural period which can be considerably large. 3 A small local peak appears in sway near the tank transverse natural period due to phase difference between the ship motion and the liquid motion. 4 No significant effect is observed on heave and pitch.

On the other hand, the liquid motion inside the tank showed significant difference between the measured and calculated forces, which suggests that slight differences in motion prediction can lead to significant differences in the liquid motion prediction. The prediction becomes especially difficult when the tank has breadth to length to ratio near 1, which is the case where both sloshing and swirling may occur. When the same mode occurred in the experiment and simulation, the force measurements were consistent.

The results show that the coupled behaviour is different when sloshing or swirling occurs. Moreover, the coupled behaviour is also dependent on the swirling rotation direction. The rotation direction did not show a constant pattern. The rotation direction is random or very hard to predict, and it strongly dependent on the initial free-surface condition.

6. ACKNOWLEDGEMENTS

Part of this research was carried out as ClassNK's Joint R&D with Industries and Academic Partners project. The authors would like to express their gratitude to ClassNK for their support.

7. REFERENCES

1. NAM B.W., KIM Y., KIM D.W., KIM Y.S., 'Experimental and Numerical Studies on Shi Motion Responses Coupled with Sloshing in Waves', Journal of Ship Research, Vol. 53, pp68-82, 2009
2. JIANG S.C., TENG B., BAI W., GOU Y., 'Numerical simulation of coupling effect between ship motion and liquid sloshing under wave action', Ocean Engineering 108, p140-154, 2015
3. HU Z.Q., WANG S.Y., CHEN G., CHAI S.H., JIN Y.T, 'The effects of LNG-tank sloshing on the global motions o FLNG system', International Journal of Naval Architecture and Ocean Engineering 9, p114-125, 2017
4. FALTISEN, O.M., TIMOKHA A., 'Nonlinear sloshing in a spherical tank', Conference Proceedings, OMAE, 2013
5. ARAI M., CHENG L.Y., WANG X., OKAMOTO N., HATA R., GUSTAVO M.K., 'Sloshing and swirling behavior of liquid in a spherical LNG tank', Proceedings of International Symposium on Ships and Other Floating Structures PRADS, 2016
6. CHEN B.F., WU C.H., 'Effects of excitation angle and coupled heave–surge–sway motion on fluid sloshing in a three-dimensional tank', Journal of Marine Science and Technology, Vol. 16, Issue 1, pp. 22–50, 2011
7. KARUKA G.M., ARAI M., YOSHIDA T., CHENG L.Y., ANDO H., 'Swirling effects and occurrence in chamfered prismatic tanks with partially filled loaded condition', Journal of the Japan Society of Naval Architects and Ocean Engineers., 2017
8. ARAI M., CHENG, L. Y., KUMANO, A., MIYAMOTO, T., A., 'Technique for Stable Numerical Computation of Hydrodynamic Impact Pressure in Sloshing Simulation', Journal of the Society of Naval Architects, Japan, Vol. 191, pp. 299-307, 2002
9. CHENG, L. Y., ARAI, M., 'A 3D Numerical Method for Assessment of Impact Loads Due to Sloshing in Liquid Cargo Tanks', Proceedings of the Fifteenth International Offshore and Polar Engineering Conference, pp. 214-221, 2005
10. ABRAMSON H. M., BASS R. L., FALTINSEN O.M., OLSEN H.A., 'Liquid Slosh in LNG Carriers, 10th Naval Hydro-Dynamics Symposium', pp.21-38, 1974

8. AUTHORS BIOGRAPHY

M Arai is a Professor at Yokohama National University. His previous research topics include Study on 3D Sloshing Simulation and Study on an anti-sloshing device for the tanks of LNG carrier and FLNG.

G M Karuka is a doctoral course student at Yokohama National University. His previous experience includes the occurrence of swirling in prismatic tanks.

H ANDO is a Senior General Manager at Monohakobi Technology Institute.

THE PRACTICAL IMPACT OF THE 2016 IGC CODE MODIFICATIONS ON A VLGC HULL DESIGN

L V Anghel, **I Hristache** and **M Negraia**, International Contract Engineering (ICE), Romania

SUMMARY

IMO's International Code for the Construction and Equipment of Ships carrying Liquid Gases in Bulk (the IGC Code) was substantially modified in July 2016. The 2016 Code's main impact on the hull design comes from increased collision protection of cargo tanks through larger clearance between tanks and side shell. The authors describe hull design changes needed to update a ten-year-old 84,500 m^3 fully refrigerated VLGC carrier design with four IMO Type A tanks to meet the 2016 Code requirements. Their aim was to minimize the modification amount relative to the original project, in order to benefit as much as possible of the previous deployed efforts and experience gained. The IGC Code modification impact is assessed, through the various structural analyses performed, from the initial scantling evaluation to the direct global and local strength analysis. The hull design changes also involved optimization of the hull form to minimize ship resistance while retaining the cargo capacity. The resulting power - speed prediction analysis shows that both the target hydrodynamic performance of the optimized hull and the stringent new 2016 IGC Code requirements will be met. Overall, the paper offers a comprehensive guide to a practical approach to VLGC hull design meeting the 2016 IGC Code.

NOMENCLATURE

ABS	American Bureau of Shipping
CoG	Center of Gravity
CFD	Computational Fluid Dynamics
ESD	Emergency Shut Down
GMf	Fluid Metacentric Height
IACS	International Association of Classification Societies
IMO	International Maritime Organization
ICE	International Contract Engineering
ICT	Independent Cargo Tanks
ISE	Initial Scantling Evaluation
KG	Reference height from keel to CoG
LGC	Liquefied Gas Carriers
LPG	Liquefied Petroleum Gas
NK	Nippon Kaiji Kyokai (ClassNK)
DNVGL	Det Norske Veritas
TSA	Total Strength Assessment
VLGC	Very Large Gas Carrier

1. INTRODUCTION

The global liquefied petroleum gas market shows good signs of improvement, due to the opening of the enlarged Panama Canal in June 2016 and increased production of shale gas in the U.S. This has allowed VLGCs to directly go through the expanded Canal to Asia, making it possible for North America's exports to become competitive with those from the Middle East.

In 2010, ICE developed a design for an 84,500 m^3 Fully Refrigerated Very Large Gas Carrier (VLGC) for Sasebo Heavy Industries Co., Ltd. The Vessel, with four (4) independent IMO Type A tanks, was designed for carrying liquefied petroleum gas / ammonia, with a 2G hazard rating. The self supporting, independent, cargo tanks were designed for a minimum temperature of – 48°C, maximum vapour pressure (IMO) of 24.5 kPa (0.25 barG) and maximum specific gravity of 682 kg/m^3 (ammonia).

The cargo tanks were of prismatic type, supported on insulation-bearing blocks and restrained by anti-roll, anti-pitch and anti-flotation chocks. To reduce the free surface effect and to increase the stability, the cargo tanks were divided by a centerline bulkhead and had a typical sloped shape in the upper part. In transverse direction, a swash bulkhead was provided at mid cargo tank length.

In January 2016, a substantially modified IMO's International Code for the Construction and Equipment of Ships carrying Liquid Gases in Bulk (the IGC Code) entered into force and the code became effective in July 2016.

In view of the expected market demand for additional VLGCs, ICE Group management instructed in 2017 the engineering team involved in the previous ICE's project to design a new VLGC vessel, of comparable size (84,000 m^3) complying with the revised IGC Code requirements. The aim of the designers was to minimize the modification amount relative to the former project, to take advantage, as much as possible, of the previous efforts put into obtaining a sound design and of the experience gained. This paper presents some of the conclusions drawn from this project.

2. IGC CODE MODIFICATIONS

Some of the most significant 2016 IGC Code changes, viewed mainly from a structural perspective, are discussed below.

2.1 CARGO TANKS LOCATION

The 2016 IGC Code modification with the biggest impact over ICE's former 84,500 m^3 VLGC design comes from the location of the cargo tanks. In essence, more protective space shall be ensured between an ICT and the side shell, as detailed below.

According to Ch. 2.6.1.2 of the 2006 IGC Code, for hazard rating type 2G ships, the cargo tanks should not be located less than 760 mm from the side shell plating. An additional recommendation has been added in the 2016 IGC Code, in Ch.2.1.1: "the proximity of the cargo tanks to the ship's shell shall be dependent upon the volume of the cargo tank". Thus, according to Ch.2.4.1 of the 2016 IGC Code, the distance between cargo tank and side shell shall increase to a required space of 1.715 m, when particularized to the mid Independent Cargo Tank of ICE's VLGC. The actual distance has been adopted as 1750 mm for the mid tank, in the revised design (see Figure 1).

In the original design, this distance was adopted as 1300 mm. This value was based on "access to spaces in cargo area" limitations, coming from IACS requirements, which were more stringent than the (760 mm) limitation imposed by the 2006 IGC Code.

To implement the 2016 ICT-to-shell minimum distance requirement, from the IGC Code, two alternatives have been evaluated with the goal to minimize the modifications over the previous design:

Variant A – consists of leaving the ICT unchanged and to modify the vessel's main dimensions, starting with the increase of the hull's breadth. This would mean to change the outer shell shape. This modification has a big impact over the speed-power estimation, hydrodynamic calculations, structural calculations and all related documents.

Variant B – consists of the reduction of the ICT breadth, while the hull's breadth remains unchanged. This preserves the vessel's hull shape below the waterline. To compensate for the loss of ICT's volume due to its breadth reduction, the tank height has been increased. Consequently, the hull depth has been equally increased. Despite this depth increase, if the draught is not changed significantly, the impact on the hull hydrostatics, hydrodynamics and even scantlings is less significant than in variant A.

In conclusion, the variant B path was selected, as shown in Figure 1.

2.2 VAPOUR PRESSURE

According to Ch.4.13.2 of the revised IGC Code - a vapour pressure P_h higher than P_0 may be accepted for site specific conditions (harbour or other locations), where dynamic loads are reduced. This is a logical concession of which designers could take advantage even before, although not stated specifically in the IGC.

2.3 ACCIDENTAL LOADS

Accidental loads requirements are provided in the revised IGC Guide (IMO, 2016), for Collision, Flooding and Fire.

Collision loads are stipulated in Ch. 4.15.1, "with an inertial force corresponding to 0.5 g in the forward direction and 0.25 g in the aft direction". This requirement has no correspondence in the 2006 IGC Code (IMO, 2006), while it has a big impact over the dimensioning of the anti-pitch chocks and ICT end bulkheads structure. However, the collision loads were included in ICE's original design, being stipulated in Section 4.3 of the ABS LGC Guide, (both for 2010 (ABS, 2010) and 2017 (ABS, 2017) editions), as Load Cases 13 and 14.

Loads due to flooding are included in Ch. 4.15.2 of the revised IGC Code. "For independent tanks, loads caused by the buoyancy of an empty tank in a hold space flooded to the summer load draught shall be considered in the design of the anti-flotation chocks and the supporting hull structure." This requirement has no correspondence in the 2006 IGC Code. However, it was considered in ICE's original design, being stipulated in Section 4.3 of the ABS LGC Guide (ABS, 2010), as Load Case 12 of Table 2.

According to the revised IGC rules, Ch. 4.3.4, the cargo containment systems shall sustain without rupture the rise in internal cargo tank pressure with no more than 20% above the MARVS (maximum allowable relief valve setting of a cargo tank), due to fire incidents. In the 2017 edition of the ABS LGC Guide (ABS, 2017), a related requirement was included in Section 3, Ch. 11.7. Additional limit state calculations have to be employed to ensure this requirement is fulfilled.

2.4 ACCELERATIONS

The formulae for vertical acceleration (used to calculate the internal pressures for ICT), given in Ch. 4.28.2.1 of the 2016 Revised IGC Code, has been changed in comparison with the one from the 2006 edition, by adding a new member, marked in red, in equation below:

$$a_z = \pm a_0 \sqrt{1 + \left(5.3 - \frac{45}{L_0}\right)^2 \left(\frac{x}{L_0} + 0.05\right)^2 \left(\frac{0.6}{C_B}\right)^{1.5} + \left(\frac{0.6 y K^{1.5}}{B}\right)^2}$$

However, since the hull main particulars (L, B, Cb) and the speed are not changed, in the discussed case, a_z increases with less than 1%.

2.5 CONSTRUCTION AND INSTALLATION

According to the revised IGC rules, Ch.4.13.7 "Loads or conditions associated with construction and installation (e.g., lifting) are to be considered". This requirement has no correspondence in the 2006 IGC Code. However, this can be considered later in the design, depending also on yard's facilities for construction and installation. Local reinforcements are generally provided for the installation loads. Generally this is not a dimensioning requirement for the overall structure.

2.6 STRUCTURAL SIMPLIFIED APPROACH

The new structural analysis requirements from the revised IGC Code have been incorporated in the updated ABS LGC Guide (ABS, 2017). However, the impact over the previous design is not significant with regard to scantlings. Part of the updated ABS LGC Guide requirements addresses simplified methods of structural design. For example, in Section 6 – Acceptance Criteria of the ABS Guide, Ch.5.6 has been added. It includes the requirements of Ch.4.17.3 of the revised IGC Code. However, these requirements refer to static and dynamic stresses when they are calculated separately. This is not the case with the applied top – down Total Strength Analysis performed by ICE, going from global coarse mesh cargo hold calculations to fine mesh local analyses.

The same is valid for Ch.5.7 of Section 6 – Acceptance Criteria from the ABS LGC Guide, where the requirements of the revised IGC Code Ch. 4.21.3 Ultimate design condition are included. There, the provided allowable stresses for Type A independent tanks are valid when the calculations are performed "by classical analysis procedures", which also is not the case when using ICE's Finite Element Total Strength Analysis.

2.7 FATIGUE

The fatigue design condition, defined in Ch. 4.18.2 of the revised IGC Code, is similar to the Ch. 4.4.5.6 requirement of the previous edition. It includes the loading and unloading contribution to the fatigue damage.

$$\sum \frac{n_i}{N_i} + \frac{n_{Loading}}{N_{Loading}} \leq C_w$$

Where:

n_i = number of stress cycles at each stress level during the life of the tank;

N_i = number of cycles to fracture for the respective stress level according to the Wöhler (S-N) curve;

$n_{Loading}$ = number of loading and unloading cycles during the life of the tank, not to be less than 1000. Loading and unloading cycles include a complete pressure and thermal cycle;

$N_{Loading}$ = number of cycles to fracture for the fatigue loads due to loading and unloading;

C_W = maximum allowable cumulative fatigue damage ratio.

It has to be noted that in the 2010 ABS LGC Guide (ABS, 2010), the low cycle contribution to the fatigue damage is not accounted for. Moreover, in the 2017 version of the same rules (ABS, 2017), the low cycle contribution to the total fatigue is considered only in relation with Type B Independent Tanks. However, the low cycle contribution to fatigue is not difficult to be assessed in meaningful locations, depending on one stress range (resulting from a pair of full-empty tanks combinations) and a predetermined number of cycles (which should be not less than 1000).

2.8 OTHER IGC CODE MODIFICATIONS

Although not directly linked to hull design, which is the topic of this paper, it is worth mentioning that there are some other significant 2016 IGC Code modifications, mainly related to safety, which should not be overlooked. Thus, for example, expanded requirements for ESD system have been provided, now including the requirement to fit a ship-shore link, instrumentation, relief valve emergency isolation and the need for comprehensive Cargo System Operations manuals. Regarding the Electrical Installations, the new requirements stress the goal of minimizing the risk of fire and explosion and introduce the definitions of Zone 0, Zone 1, Zone 2, and Non-hazardous area, instead of "gas safe" or "gas dangerous", etc.

3. IGC CODE MODIFICATION IMPACT

3.1 ISE - GENERAL CONSIDERATIONS

To keep the consistency and to be able to properly evaluate the real impact of the IGC Code modifications, the comparison between the two VLGC designs was based on successive editions of the same Classification Society Rules (ABS), (ABS, 2010) and (ABS, 2017).

It has to be noted that, for LPGCs, different Classification Societies use different formulas for calculating the required scantlings. For example, if the NK rules are followed, a sturdier structure results for the ICT tank, although the same IGC pressures are applied. Moreover, some Classification Societies have included in their rules simplified formulas for the ICT pressure calculation, based on Tank's CoG accelerations, as an alternative to the IGC provisions. For example, DNVGL, and NK allow for the same simplified approach with regard to ICT pressures.

3.2 ISE - PARTICULARIZATION

Practically, the authors found that the 2016 IGC Code changes could be implemented with no significant midship section scantlings modifications. In the analysed case, although the cross section geometry has been changed, by ICT height and hull depth increase and ICT breadth reduction, these modifications are small, relative to the overall dimensions of the section, being in the range of 1-2%. The pressure modifications for the revised design are also around 1-2%. On the other hand, the original scantlings surpassed the requirements with a certain margin, generally higher than 5%. Thus, for example, a 5%-10% reserve in a stiffener section modulus is common, since stiffeners should adhere to a certain standardization and, thus, will not fall exactly on the rule required value.

Due to the increased hull depth, the effective hull girder section modulus rose with about 1.6%.

3.3 TSA – GENERAL CONSIDERATIONS

The scantlings obtained from the Initial Scantling Evaluation (ISE) have to be further checked by a total strength assessment (TSA). They may be confirmed or increased by the TSA but it is generally not acceptable to reduce them. For example, the TSA is generally dimensioning for the bottom plate thicknesses, which is usually increased in this stage, due to transverse load buckling, which can not be appropriately considered in ISE.

During the TSA, the structure has to be checked against the Yielding, Buckling, and Fatigue failures modes.

Actually, due to the peculiarities and complications involved, an LGC structural design is difficult to be conceived without an extensive TSA support.

In case of a Type A LGC, the departure from the treaded path followed for a conventional ship, with regard to the strength assessment, is given by the non-linear nature of its global analysis. This is a consequence of the relative cargo tanks independence, needed to minimize the heat transfer. Thus, during hull deformation, some of the inner bottom vertical supports may lose contact with the tanks. This has to be accounted for in the finite element model, implying successive analyses, in an iterative process.

To obtain a rational structural design, contact analyses may be needed for the antiflotation chocks to wing tanks structure interaction, in the accidental flooding scenario.

Friction intervenes in the interaction between the independent cargo tanks and the hull through the vertical supports. This phenomenon has an impact when accurate fatigue evaluation is of concern in the tank support area, since the friction coefficient changes, when sliding occurs.

With regard to the Total Strength Analysis, the 2016 revised IGC Code does not change the structural modelling principles or global finite element modelling routine.

For LPG type of ships, hull and cargo tank structures within and beyond 0.4L amidships need to be evaluated. This means that there are three regions to be subjected to structural assessments. The midship region is the first candidate for the TSA. This area is easier to be modelled due to its regular shape and is also subjected to the biggest hull girder bending loads.

The FEM models used in TSA have been developed starting from the ISE scantlings. They have the target cargo tank in the middle of the model and extend one half the length of the adjacent tanks fore and aft (see Figure 2), as required in (ABS, 2010).

The fore-most and aft-most cargo tanks models benefit from the experience gained during midship analysis. In addition, they reveal the effects of changes in hull form, cargo tank shape, load characteristics and particularly in tank support arrangement, completing the overall structural picture.

Due to the tapered hull forms, the number of supports has to be reduced toward the extremities of the end cargo holds, increasing the supports design loads and complexity, in these regions.

Also, with regard to the Total Strength Analysis, the 2016 revised IGC Code does not affect the boundary constraints practice, applied for local and hull girder sub load cases.

For the particular ICE analysis, the hull girder loads have not changed significantly. Tank volumes are sensibly reduced only in the forebody part (Cargo Tank 1), with about 5%. The hogging still water bending moment increases by 3% while sagging decrease by 1% due to small tank and general arrangement modifications, which move some weights toward the ends of the ship. Wave bending moment have not changed.

As already mentioned as a general remark, with reference to the local loads, the vertical acceleration formula (used to calculate the internal pressures for the ICT) has changed in the 2016 revised IGC Code. However, its modification was found to be of less than 1% in the studied case.

Moreover, the height of the ICT was increased by 300 mm. This represents less than 2% of the tank height. The modifications fall within the previously included inherent engineering margins, as commented in Chapter 3.2.

The standard ABS design load cases for yielding and buckling strength assessment and for fatigue strength assessment are not affected by the new IGC provisions. As already mentioned in Chapter 2.3, some of the additional requirements of the revised IGC Code are covered by the 2010 ABS LGC Guide (ABS, 2010), used in the original VLGC design.

3.4 TSA – PARTICULARIZATION

For the 2010 84,500 m^3 VLGC structural design, during the TSA phase, the authors analysed 3 global Finite Element models corresponding to the Aft Body, Midship and Fore Body regions, respectively (see Figure 2 and Figure 3). Figure 4 shows an overview over the stress results for the Midship FE Model.

83 critical areas, agreed with ABS, were investigated in detail, based on finer mesh (1/5 ~ 1/10 of longitudinal spacing) and 30 critical structural details (covering 211 locations) were evaluated in accordance with the fatigue strength requirements, based on plate thickness mesh size finite element models.

The final structural design of the vessel is the outcome of all these direct analyses. The whole design by analysis effort were supported by the ABS LGC version 2.0 software, provided by courtesy of the American Bureau of Shipping.

Since it is such a laborious process, the TSA phase was not repeated for the new 84,000 m^3 VLGC. However, based on the experience gained during the previous analysis, the following considerations were made.

One of the most challenging critical areas of the hull is the dome opening area in the deck (see Figure 5). There, 27mm higher tensile steel insertions have been made in deck and coaming. It was a difficult task to find the shape of the opening's corners, so as to keep the stresses under the allowable value. After many iterations, an acceptable shape, consisting of a succession of three radii, was finally decided. Even the shape of the corner's brackets, (supporting the coaming) and their face plate's geometry was important, in order to comply with the strength and fatigue criteria imposed by the rules.

In the revised configuration, shown in Figure 1, the deck is 300mm higher than in the original design. The section modulus of the hull cross section increased with about 1.6%, as a result. Therefore, the stresses are expected to decrease in the dome opening area and its configuration can, very likely, be preserved. Increasing the hull depth has to be beneficial for the dome opening in the deck.

Another critical structural area, identified from the global finite element analysis and also revealed from service experience is represented by the side frames. Their most stressed regions occur in the middle of the hold area, at the end brackets locations, especially in their lower part (see Figure 5). For the revised design of the hull, in line with the 2016 IGC Code, the side frames' span has been increased. Higher stresses should occur in their end bracket toes. These represent critical locations, which should be especially considered from fatigue point of view.

Even for the original configuration of the hull, it was a difficult task to find a suitable shape of the side frames' bracket toes, to satisfy the fatigue requirements. Soft bracket toes and tapered face plate with chamfered end were provided. Many iterations were performed by variations of the shape parameters (height of the soft toe, radii of the soft toe, bracket radius and start angle) until a satisfying configuration was found. A balance was reached between the fatigue life at the critical locations found at the end bracket toe weld, along the curved free edge of the bracket toe and at the weld between the face plate end and the web of the side frame. A more permissive S-N Curve is applicable along the free edge of the bracket toe. The modification of the side frame span would certainly impose a re-evaluation and very likely a re-design of its end brackets.

The flooded condition is dimensioning for the transverse bulkheads of the hull. According to the ABS LGC Guide, the pressure acting on the transverse bulkhead depends on the flooded draft and the imaginary freeboard depth. For the analysed design, the flooded mean drafts at midship are less than 12m, while 80% of the freeboard depth is about 18m. Thus, the freeboard depth, which was increased with 300 mm, is dimensioning. As a result, the pressure acting over the hull transverse bulkhead will show a small increase. For the determined configuration of the transverse bulkhead, the highest stresses occur in the vertical web stiffeners, in their lower part, close to the bracket toes and at the intersection with the first horizontal stringer and also at their upper bracket toes (see Figure 5). Special attention has to be given to these critical structural areas, in the revised design.

Originally, there was a big challenge to solve the fatigue problems occurring at the knuckled inner bottom to lower hopper sloping plate intersection. There, besides the full penetration welds provided and scallops elimination, transverse and longitudinal brackets of 15 mm thickness were added, to decrease the critical stress ranges. Since the geometry remains the same, the provided reinforcements, very likely, remain valid. However, minor thickness changes may be required, due to the small modifications of the loads (ballast tank height increases by 300 mm).

Two critical structural areas can be mentioned with regard to buckling:

- The hull side shell where the applied transverse framing system is more susceptible to buckling due to the longitudinal hull girder stresses. In the revised design, the span of the side panels was increased by 300 mm, which tends to make the side more vulnerable.
- The hull non watertight double bottom girders. There, due to the increased section modulus the

longitudinal stresses should decrease a little, having a beneficial effect in the buckling evaluation.

Inside the ICT, one of the most problematic supporting members was the transverse bulkhead lower stringer. High stress critical areas from yielding and fatigue point of view occur at the ends of these stringers. Also yielding and buckling problems were found in the centerline bulkhead of the ICT, where these stringers end. For the revised design, however, the transverse bulkhead's stringers span decreased by about 450 mm, which should slightly relief the stresses, despite the small pressure increase.

Since the ICT tank volumes have not been increased and since their structure will not change noticeably, the supports and chocks configuration don't need to be changed. As the height of the ICT was increased by 300mm, the CoG vertical position increased as well, by about 150mm. This represent less than 1% of the tank height. The accelerations corresponding to the CoG of the tank increase, too, but unnoticeable. On the other hand, the tank's side "cantilevers" decrease in width by 450 mm each, which tend to reduce the extra loads acting on the side vertical supports.

The heaviest loaded vertical supports are those from the corners of the ICT, where the maximum vertical reactions reached about 1200 tf (see Figure 5). This intense load transfer radiates extreme stresses in all surrounding structures. The inner bottom structure was reinforced in this area, with thicknesses increasing up to 20 mm for the affected girder panels and even up to 25 mm for some floor panels, due to shear effects. Similar plate thicknesses were required inside the ICT for the corner reinforcements (end transverse web frames in their lower corners and several connected stiffeners and brackets), to dissipate the peak loads to the surrounding structure, without surpassing the allowable shear stresses. Since the loads only increase marginally for the revised design, while some extra strength margins have been incorporated previously, no modifications are expected for the vertical supports and their surrounding structure.

For the anti-pitch chocks, the dimensioning longitudinal accelerations, corresponding to the forward and aftward collision conditions, have not been changed (0.5g forward and -0.25g aftward) therefore there should be no implications for the sturdy, boxed, anti-pitch chocks configuration.

The Anti-Roll Chocks loads should not change significantly, either. The ICT mass does not change substantially and although its centre of gravity position increases, the distance between the anti-roll chocks (placed on the inner bottom and below deck) increases twice as much, which makes the chocks reactions to remain (theoretically) almost unchanged.

4. INTACT STABILITY

The 2016 revised IGC Code has extended the stability requirements to cover all possible conditions of the vessels, as follows:

- the stability of the ship, in all seagoing conditions and during loading and unloading, shall comply with the requirements of the International Code on Intact Stability. This includes partial filling and loading and unloading at sea, when applicable
- stability during ballast water operations shall fulfil stability criteria
- the free surface effect in undamaged compartments shall be calculated by a method according to the International Code on Intact Stability

ICE aligned its VLGC concept design to fulfil the latest requirements of IGC Code.

In this regard, 116 loading conditions were generated, from which two (2) main groups are distinguished:

- Harbour Conditions
- Sailing Conditions

The revised ICE VLGC concept was improved so that no ballast water is needed between departure and arrival stages, for all fully loaded conditions.

The stability calculations have been performed by means of KG limiting curves for intact stability requirements.

All harbour conditions (applicable for sheltered areas only) have a positive GMf value ($\geq$ 0.15 m); the KG limiting curves are not applicable to these conditions.

All sailing conditions comply with the intact stability criteria. Even the nearest loading condition (FL Ammonia+100% Bunkers) relative to the KG limit curves has a significant stability reserve, which leads to the conclusion that the vessel has a good stability in intact state for all the loading conditions considered.

The complete overview of the KG limiting curves and loading conditions are presented in Figure 6.

5. DAMAGE CONDITION

With regard to the survival requirements, the revised IGC code included the shifting range principle, according to MSC/Circ.406/Rev.1, where "the 20° range may be measured from any angle commencing between the position of equilibrium and the angle of 25° (or 30° if no deck immersion occurs)".

This damage condition was investigated for the revised 2017 ICE VLGC design.

The conclusion of the analysis is that the ship meets the applicable requirements regarding the protective location of cargo tanks and the survival capability regarding the damage stability criteria.

The most severe combination between an intact loading condition and a damage scenario is the combination between the loading condition with Ammonia+100% bunkers with a side damage scenario, when 9 compartments are involved. See figure 7.

The calculations were performed using the AVEVA Marine ID 12.1.SP4 software.

6. CFD HULL FORM OPTIMIZATION

6.1 CFD APPROACH

Concurrently with the implementation of the latest IGC Code requirements in the revised design, a hull form optimization was performed.

In order to optimize the VLGC hull shape, from resistance point of view, systematic calculations were carried out, employing the powerful CFD code SHIPFLOW 6.2. The calculations were performed for an 11.4m design draft with no trim.

6.2 HULL MODIFICATIONS

Fifteen (15) versions of hull were analysed and compared from the hydrodynamic performance point of view. Successive runs were carried out to determine the distributions of the main hydrodynamic parameters quantifying the flow around the hull.

A brief view over the hull modifications operated between the initial hull version - V0 (plot in red) and the final hull version - V15 (plot in blue) is presented in Figure 8.

Hull modifications were conducted in two areas – aft hull region and fore hull region.

6.2 (a) Aft hull region modifications

In the aft hull region, the shape modifications consist in:

- raising of Stern profile to increase the distance from propeller up to the hull;
- rounding the bilge keel to avoid the vortices appearance. In this respect the FOB (flat of bottom curve) has been modified;
- Design a small aft bulb below shaft line to improve the wake in disk propeller.

6.2 (b) Fore hull region modifications

In the fore hull region, the shape modifications consist of:

- Changing the bulbous bow (form and dimensions);
- Rounding the bilge keel to modify the pressure coefficient on the hull.

6.3 CFD RESULTS

The ship's forms have been shaped after a number of successive runs, targeted to improve the hydrodynamic parameters which quantify the flow around the hull, for a target speed $V_S = 17.25$ Kn.

The implemented modifications improved the hydrodynamic behaviour of the hull, mainly by reducing the wave resistance coefficient (C_W).

The overall ship resistance coefficient of the best performing hull shape (V15) has been improved by about 3.25% when compared to the initial hull shape (V0).

In terms of ship wave resistance coefficient, the final hull variant offers an improvement of about 14% over the initial hull variant.

Figure 9 presents a wave profile comparison between the two extreme hull variants. A more settled wave profile, with lower peaks, corresponds to the final hull variant.

Figure 10 presents, by comparison, the improvement of the wake distribution in the disk propeller as a result of the small aft bulb provided below the shaft line, in the final hull shape design.

These results, in conjunction with data from the "Speed Power prediction", gives confidence in reaching the target ship's performance. They have to be further confirmed in towing tank tests made on the ship's model.

For the power speed prediction of the ship, the AVEVA ID module was used.

The calculation of the bare hull resistance is based on the Holtrop & Mennen method (Holtrop & Mennen, 1982), a statistical regression method with an accuracy estimated to around 10-15%. The power–speed prediction was also verified using the results from the CFD simulation with the SHIPFLOW software. A 7.4 m diameter ship propeller was found to be optimum.

7. CONCLUSIONS

In 2017, an engineering team from ICE was challenged to update their former 2010 VLGC design to comply with the IGC Code requirements revised in 2016. The aim of the designers was to minimize the modification amount relative to the former project and to benefit, as much as possible, of the previous efforts put into obtaining a sound design.

To keep the consistency and to be able to properly evaluate the real impact of the IGC Code modifications, the comparison between the two LGC structural dimensions was based on successive editions of the same Classification Society Rules (ABS).

The 2016 IGC Code modification with the biggest impact on a former LPG design comes from the location of the cargo tanks. In essence, the new code requires more protective space between ICT and side shell. To implement this requirement, two alternatives were evaluated, namely to leave the ICT unchanged and to modify vessel's breadth or to reduce the ICT width while the hull breadth remains unchanged. Although the depth of the hull had to be increased, the second alternative was selected.

The authors found that the other IGC Code modifications did not change significantly the original structural dimensions. In the analysed case, the overall strength requirements increase with around 2-3%, well within the margins provided in the original design.

Since a full TSA is an extremely laborious and complex process, involving non-linear, contact and friction problems, it was not performed for the revised design. However, an evaluative review was performed, revealing some areas of concern (like side shell area), mainly due to the hull depth increase.

Intact stability and damage conditions evaluations were performed, having results that comply with the revised IGC Code requirements.

A hull form optimization was also carried out, using the CFD approach, resulting in a 3.25% improvement of the overall ship resistance coefficient. The power - speed prediction analysis showed that the target hydrodynamic performance of the optimized hull could be improved, while the new 2016 IGC Code requirements will be met.

8. REFERENCES

1. IMO, 'International Code for the Construction and Equipment of Ships Carrying Liquefied Gases in Bulk', *IMO Publication*, 2006.
2. IMO, 'International Code for the Construction and Equipment of Ships Carrying Liquefied Gases in Bulk', *IMO Publication*, 2016.
3. ABS, 'Guide for Building and Classing Liquefied Gas Carriers with Independent Tanks, Hull Structural Design and Analysis', 2010.
4. ABS, 'Guide for Building and Classing Liquefied Gas Carriers with Independent Tanks, Hull Structural Design and Analysis', 2017.
5. HOLTROP J., and MENNEN G.G.J., 'An Approximate Power Prediction Method', *International Shipbuilding Progress, 29, 335, 166–170,* 1982.

AUTHORS' BIOGRAPHY

Dr. Lucian Viorel Anghel holds the position of Senior Engineer in the Naval Architecture & Structure department of ICE. He received his MSc. degree in Naval Architecture and Marine Engineering from the Dunarea de Jos University and was awarded his PhD degree in 2002. He was a member of the Structural Analysis team which developed ICE's 2010 VLGC design and has also been involved in LNGC to FSRU conversion efforts.

Ionel Hristache is a Principal Engineer in the Concepts Development of ICE. He received his MSc. degree in Naval Architecture and Marine Engineering from the Dunarea de Jos University in 2011. He was the team leader of Initial Design Team which developed ICE's 2017 VLGC concept design and has also been involved in a number of ship design and ship conversion projects.

Marcel Negraia has an MSc degree in Naval Architecture from the Dunarea de Jos University in Galati, Romania. He worked in increasingly responsible positions in the Damen Group, until he joined the ICE Group as a Project Manager. As such, he has been responsible for design of drill ships, gas carriers and other complex ships for clients around the world. In 2017 he was appointed as Project Manager for developing ICE's new VLGC design.

APPENDICES

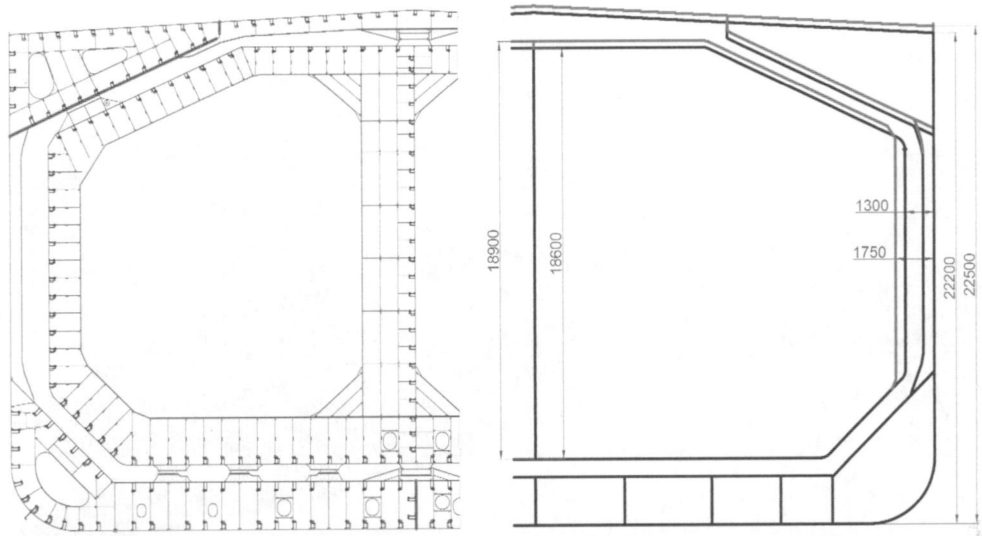

Figure 1. Midship cross sections. Modifications triggered by the 2016 IGC Code Rules (IMO, 2016)

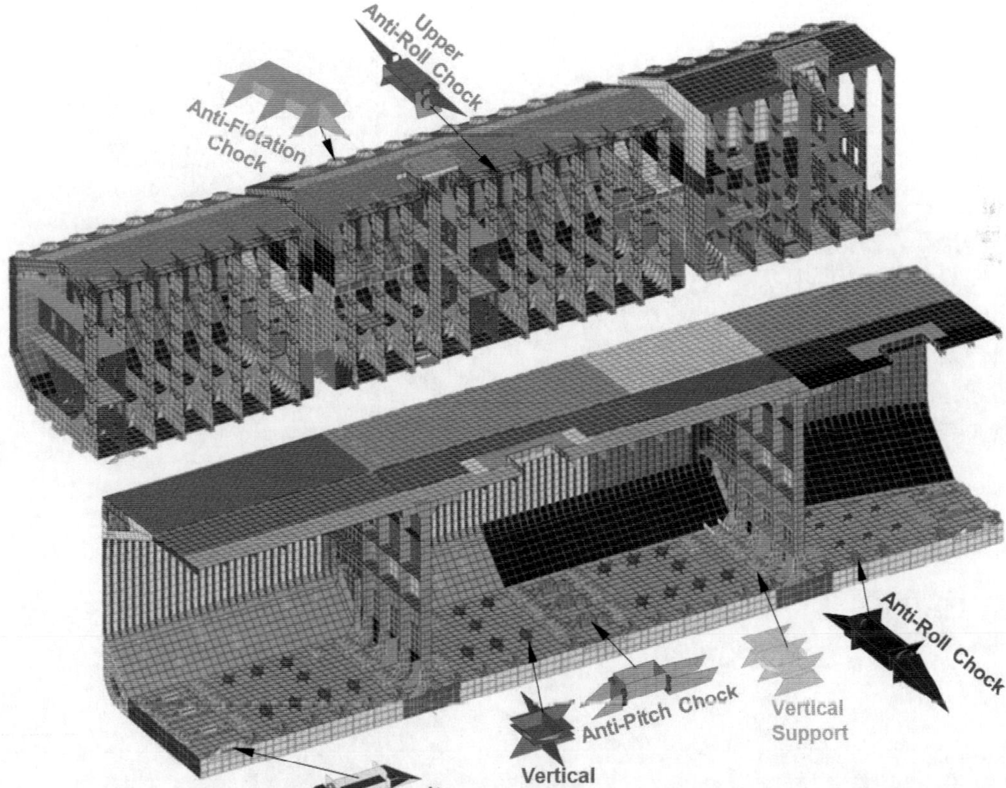

Figure 2. Midship Global FE Model of Hull and Independent Cargo Tanks. Supports and Chocks FE details.

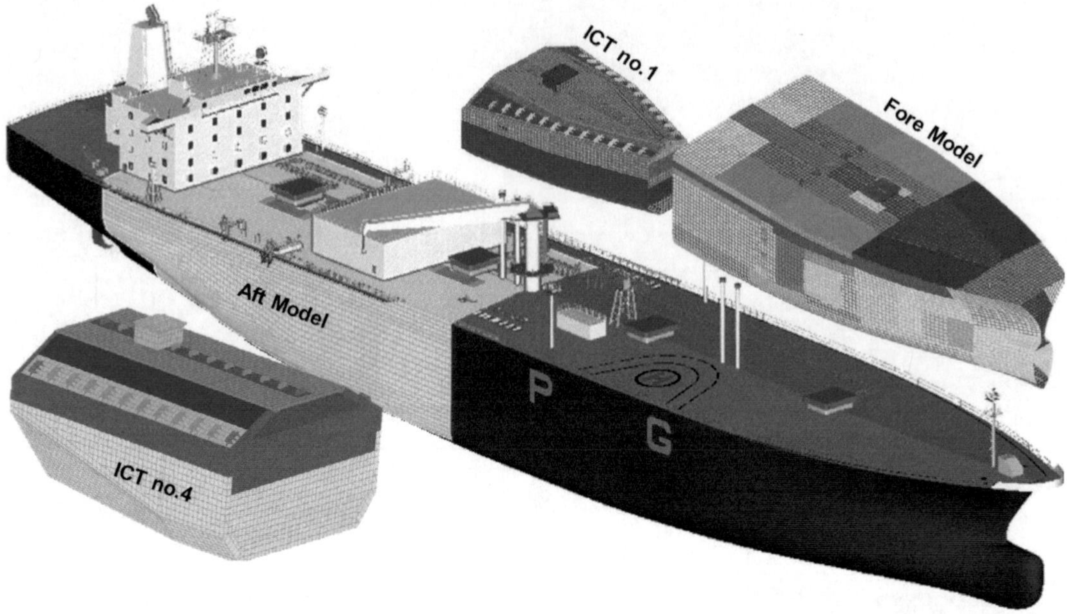

Figure 3. Hull and Independent Cargo Tanks of Aft and ForeBody Global FE Models. Plate thickness related colours.

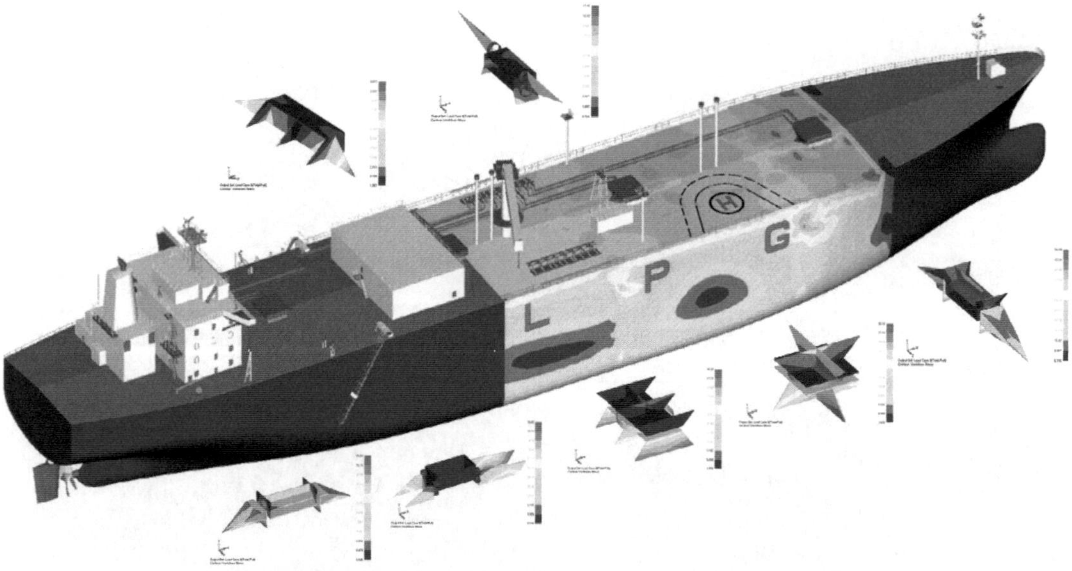

Figure 4. Stress Results for Midship Global FE Model. Details with supports and chocks stresses, at coarse mesh size.

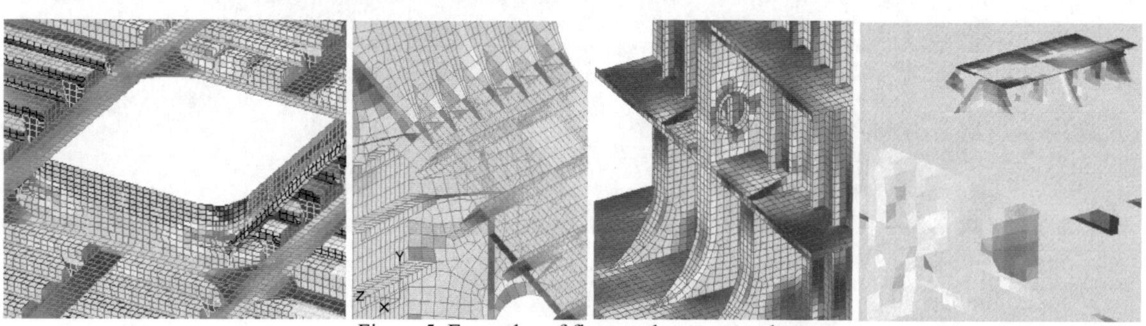

Figure 5. Examples of fine mesh stress results

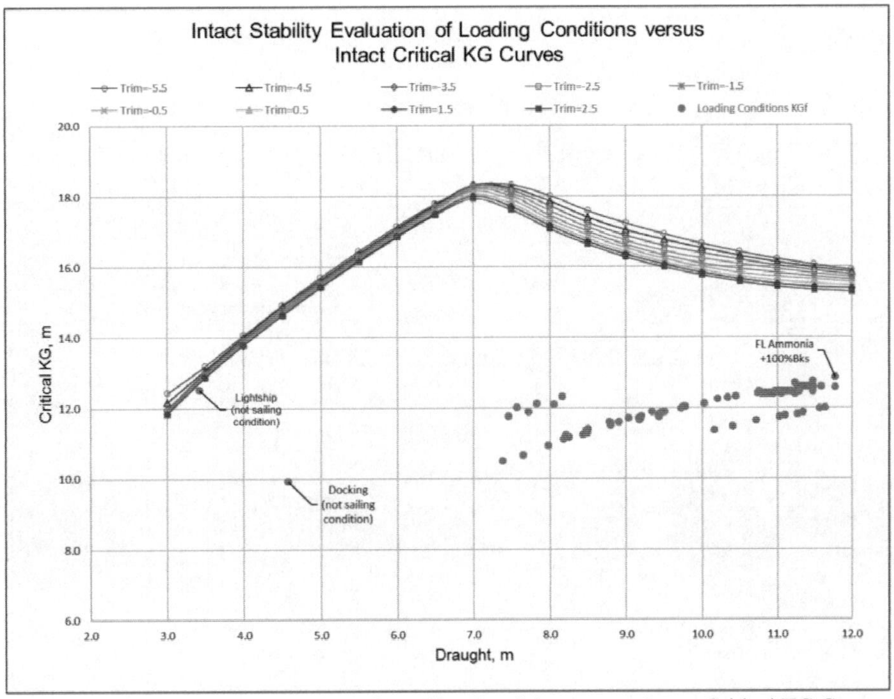

Figure 6. Intact Stability Evaluation of Loading Conditions versus Intact Critical KG Curves

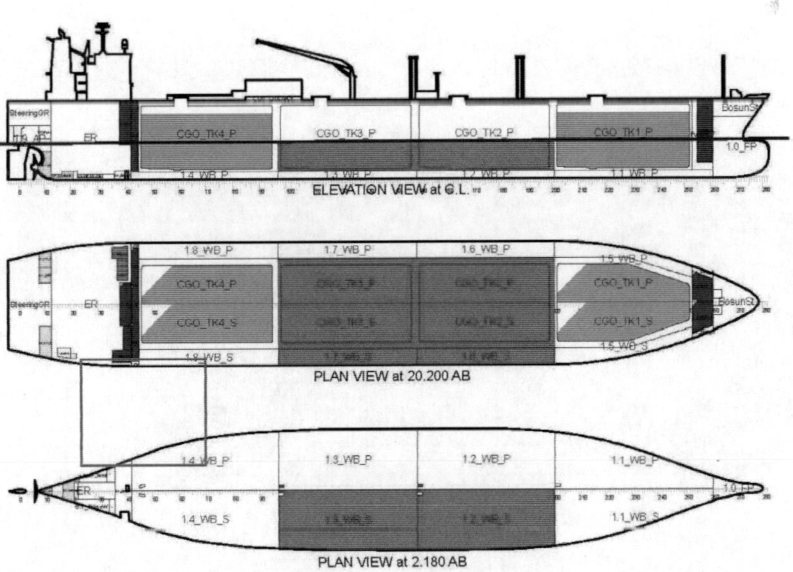

Figure 7. The most severe loading condition in a damage scenario

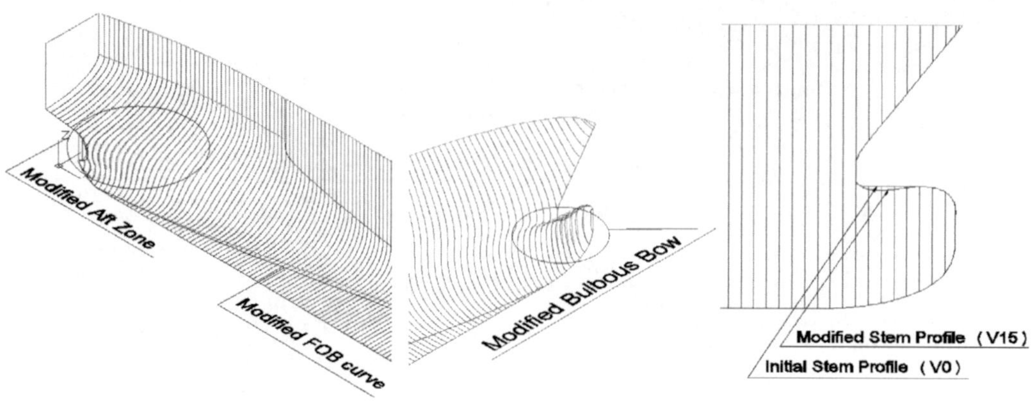

Figure 8. Hull Modifications at the Aft and Fore Ends

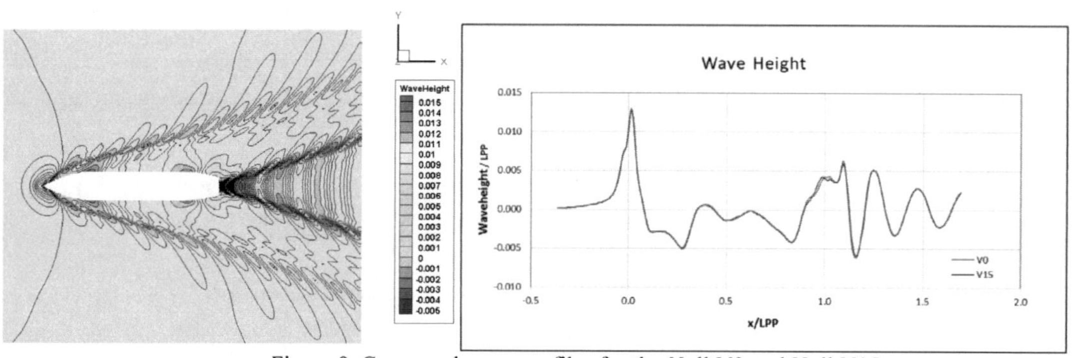

Figure 9. Compared wave profiles for the Hull V0 and Hull V15

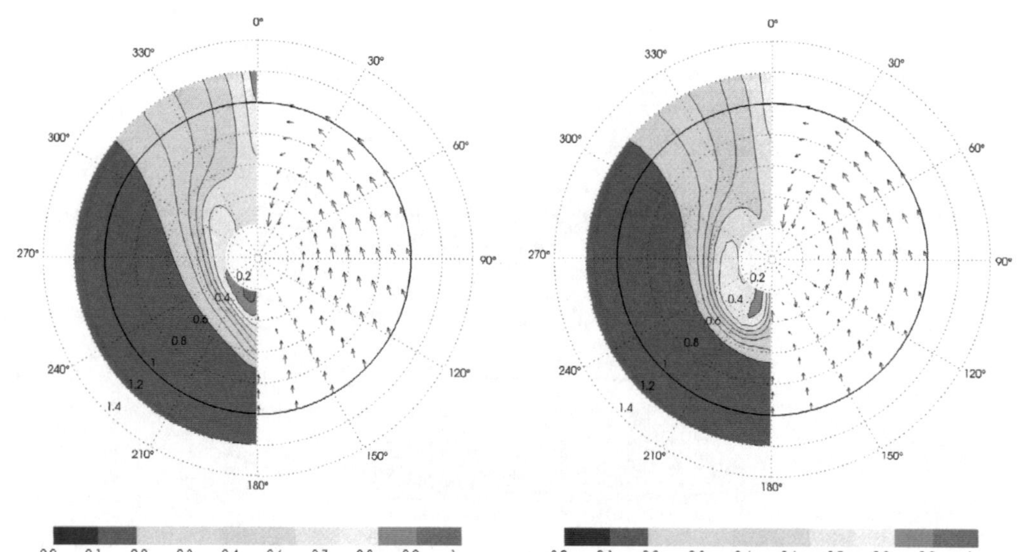

Figure 10. Wake distribution / Velocity vectors

AUTHORS' CONTACT DETAILS

George Mermiris
Tritec Marine Limited, UK
George.memiris@tritec-marine.com

Carlos Guerrero
Bureau Veritas Marine & Offshore, France
carlos.guerrero@bureauveritas.com

Robert Tustin
Lloyd's Register, EMEA, UK
robert.tustin@lr.org

Keith Hutchinson
Babcock International,UK
Keith.w.hutchinson@babcockinternational.com

Thomas Lamb
University of Michigan, USA
nalamb@umich.edu

Antonis Trakakis
Arista Shipping, Greece,
trakakis@aristashipping.com

Mélodie Noris
GTT, France
mnoris@gtt.fr

Effiong Ekanem Attah
University College London, UK
e.attah.12@ucl.ac.uk

B O Bernstsen & K Høglund
Høglund Marine Automation AS, Norway
Bengt-olav@hma.no

George Dimopoulos
DNV GL, Greece
george.dimopoulos@dnvgl.com

Jonathan Strachan
Houlder Ltd, UK
Jonathan.Strachan@HoulderLtd.com

Makato Arai
Yokohama National University, Japan
arai.ynu@gmail.com

Lucian Anghel
Ice Contract Engineering Ltd
lucian.anghel@icepronav.ro

The Royal Institution of Naval Architects
8-9 Northumberland Street
London, WC2N 5DA
United Kingdom

ISBN 978-1-5108-8306-2